# AMONG THE
# BONE EATERS

**ANIMALIBUS** VOL. 8
OF ANIMALS AND CULTURES

Nigel Rothfels
*General Editor*

ADVISORY BOARD:
Steve Baker *(University of Central Lancashire)*
Susan McHugh *(University of New England)*
Garry Marvin *(Roehampton University)*
Kari Weil *(Wesleyan University)*

*Books in the Animalibus series share a fascination
with the status and the role of animals in human
life. Crossing the humanities and the social sciences
to include work in history, anthropology, social and
cultural geography, environmental studies, and liter-
ary and art criticism, these books ask what thinking
about nonhuman animals can teach us about human
cultures, about what it means to be human, and
about how that meaning might shift across times
and places.*

# AMONG THE BONE EATERS

## ENCOUNTERS WITH HYENAS IN HARAR

### MARCUS BAYNES-ROCK

#### WITH A FOREWORD BY ELIZABETH MARSHALL THOMAS

THE PENNSYLVANIA STATE UNIVERSITY PRESS

UNIVERSITY PARK, PENNSYLVANIA

Typeset by
Regina Starace

Printed and bound by
Sheridan Books

Composed in
Dante MT

Printed on
55# Natures Natural/70# Matte

Bound in
Arrestox Butterscotch and
Multicolor Cordovan

Library of Congress Cataloging-in-Publication Data

Baynes-Rock, Marcus, 1965– , author.
Among the bone eaters : encounters with hyenas
in Harar / Marcus Baynes-Rock ; foreword by
Elizabeth Marshall Thomas.
        pages            cm — (Animalibus)
Summary: "An interdisciplinary study of the relationship between hyenas and the people of Harar,
Ethiopia"—Provided by publisher.
Includes bibliographical references and index.
ISBN 978-0-271-06720-9 (cloth : alk. paper)
1. Hyenas—Behavior—Ethiopia—Harer.
2. Hyenas—Effect of human beings
on—Ethiopia—Harer.
3. Human-animal relationships—Ethiopia—Harer.
4. Harari (African people)—Ethiopia—Harer—Social
life and customs.
5. Harer (Ethiopia)—Social life and customs.
I. Title.  II. Series: Animalibus.

QL737.C24B39 2015
599.74'3—dc23
2015011563

FOR BABY

# CONTENTS

# FOREWORD

Ever since the Neolithic, we have left no stone unturned in distancing ourselves from other species, modifying our language, our laws, and our thoughts accordingly. A person is either *he* or *she*, an animal is *it*, we have thoughts, they have instincts, we have language, they have vocalizations, we have culture, they just run around in the woods, doing whatever it is they do, and so forth. During this time, anthropology was invented, as was zoology, and both had no choice but to continue the traditions. We mammals have evolved more than five thousand species, and zoology is concerned with all but one.

The responsibility for studying and writing about that last remaining species—and only that one—has framed the nature of anthropological research in a negative way. When anthropologists study cultures that involve animals—the cultures of hunter-gatherers, for instance—little notice is taken of the animal side, although the animals may strongly influence the people's lives, not simply by existing but by behaviors they have chosen to adopt.

Animals such as birds and mammals are as sentient as we are, with more on their minds than just eating, mating, and fighting. We seldom recognize this, especially if we think that "culture" means poetry, music, and art. But the fact is that groups of the same species living in different areas behave in different ways, strongly suggesting that more than mere instinct is directing them. We're speaking now of wild animals, who adjust their customs to conform to the circumstances in which they find themselves, and when groups of such animals become involved with people, the cultural features of both participants adjust. If anthropologists arrive to study the people, their studies mention the animals, of course, but only because of their value to the humans or as they occur in the mythology. When the economics and the myths have been recorded, the job is done.

The reason I have strong feelings about this problem is that I'm guilty of promoting it myself, something I would never have done had I known more than I did. I was nineteen at the time, so now that I'm a senior citizen I can forgive myself a little, because I've learned a few things since, but the fact remains that a tremendous opportunity was wasted and can never be recovered because the cultures in question have vanished and will not be seen again.

In the 1950s I participated in an anthropological study of !Kung-speaking Bushmen in the Kalahari Desert, now part of Namibia and Botswana. The Bushmen when we met them were pre-contact—members of our group were essentially the first non-Bushmen to penetrate the 250,000 square miles that formed their area (then generally known to the European population as "the end of the earth"), and we gathered much information about the people and their way of life. In that dry savannah, the only water was in certain waterholes, each of which supplied a Bushman encampment and a pride of lions, and, interestingly, the two groups didn't interfere much with each other. Leopards and hyenas lived there too, and sometimes hunted the people, but the lions did not.

They seemed to have a truce with the Bushmen. Over perhaps a hundred-year period, among a population of maybe two thousand people, none of whom had weapons that could kill a lion in a timely manner, and all of whom spent their days finding food in the veldt and slept on the ground in or near small, dome-shaped grass shelters, a study suggested that only one person was killed by a lion, and this was a special case because the person was a paralyzed girl who dragged herself along the ground. The normal perceptions that lions would have had about our species probably did not apply to her. Otherwise, the human-lion truce could have been as old as thirty-five thousand years, which archaeologists later determined to have been the continuous occupancy of at least one Bushman encampment.

Did we explore this fascinating situation? We did not. We were there to do anthropology, so when it came to lions, the information we collected was the Bushmen's lion lore. The lore told us much about the people, and we were satisfied with that. Had we known better, we would have seen that it said much about lions too. But at the time, non-Bushmen knew relatively little about lions, so we didn't see what the Bushman lore reflected.

As it turns out, lions elsewhere often hunted people. It was just the Kalahari lions who did not. Why this was so may never be understood, but their restraint was ancient and of course there was a reason—a good one, which will never be learned because after both groups came in contact with our so-called civilization, the Bushmen acquired livestock, which attracted the lions, and white hunter-guides made money from American and European sportsmen who shot the lions, all of which has changed the lions' opinion of our species. Today, the Kalahari lions—those who are left or who moved there from elsewhere—hunt whatever they can.

This is not to say that we didn't notice the lions. Many times, when we camped, the local lions would come to look at us. But we slept on the ground without a worry, even after we woke one morning to find lion tracks right next to our sleeping bags. The lions must have been standing right over us looking down at our faces. More often, we would hear lions roaring or would see their eyes shining in the firelight and realize that several lions had come to look at us. From these observations they knew more about us than we knew about them, because we never watched them for any length of time, or tried to recognize them as individuals or record the time of night when we saw them, or note where the moon was when they came, although, as it happens, moonlight or lack of it has a profound influence on what lions decide to do. We would have learned a great deal if we had been more observant. The only redeeming feature is that lions are so compelling that one remembers every lion that one has ever seen. This has helped me to reconstruct some aspects of the interesting Bushman-lion truce, but that's not the same as a study.

The zoologists don't do much better when it comes to studying human-animal interactions. Normally, wildlife biologists conduct their studies in game parks such as the Serengeti, environments that the animals have all to themselves because the humans stay in their cars. Not even the scientists have contact with their subjects. They watch them from their vehicles, using photographs to identify individuals and radio collars to learn where they go. Yet not so long ago, when the Serengeti was just wild land like any other such area, when the people were not tourists but hunter-gatherers—the hominin members of the indigenous fauna—human-animal interactions took place all the time, and their cultures adapted because of them. How helpful it would be if some of these interactions had been studied!

Enter *Among the Bone Eaters*, a flagship book that breaks with all tradition. If not for Marcus Baynes-Rock, who is every inch a scientist, the complexity of human-hyena relations might never have been revealed. When exposed to a human culture that mixed with a hyena culture, he opened a whole new world, not only to anthropology but also to zoology, because he studied both cultures together. I believe that he is the first scientist ever to treat a group of individuals, some of whom were people and others hyenas, as if they were parts of a single, mixed culture, the many-layered culture of Harar, the town he studied in Ethiopia.

And why is this important? Because when it comes to relationships and mixtures of lifestyles, probably not a single life-form on the planet is limited to its own kind.

We are not speaking here of people and their pets, although some people have tigers and lions as pets, often with great mutual attachment. We're speaking of self-reliant, independent animals mingling with humans for the benefit of both. As we look around us, we see that partnerships of every kind are flourishing—trees and fungi, wolves and ravens, ants and aphids, ratels and honeyguides, badgers and coyotes, and many more. Even bacteria have mixed-species relationships, probably more types than those of any other domain. Bacteria don't need to make cultural adjustments in their interaction with others, but in forming their symbiotic relationships they started a custom that is now echoed in all social animals and in some not so social animals as well. If we're social, we need to find ways of dealing with our own kind, and if another species is somehow introduced, we need to deal with them too.

In Harar, the people fed hyenas as a tourist attraction, the complexity of which this book describes, so the hyenas gained food and the people gained money, a type of mutual benefit that is far from unique in the natural world. The Harar hyenas were wild, and although they came to Harar for food, they also foraged on their own, hunting in the bush around the town. They had lost nothing of their original culture, which would have included attacking and possibly eating some of the people. As for the people, they would rightfully have feared hyenas. But in Harar, each species came to understand much about the other and modified their earlier views. This was possible because both species had modified their view of the other. Hyenas are knowable if one has reason to learn about them.

One envies Marcus as he sits in the bush with a hyena beside him while they savor the night. One envies him as he follows a hyena on the network of paths made by humans and hyenas in and around Harar, the hyena deciding which paths to take. "Envies" might be the wrong word here—"admires" might be better—but I can't deny a twitch of envy because many years ago when I was in the Kalahari I almost had a similar experience—a brief encounter with a certain hyena. We saw hyenas often enough—they came into the encampment. Now and then one would poke his head through the flap of my tent to see if I was awake and healthy, and finding that I was he would withdraw. But once I was walking alone in the veldt and came around the side of a hill to see, not thirty feet away, a big hyena, I think a female, standing in the opening of her den. We both stood still and looked at each other. She seemed a bit stiff but she didn't threaten me. I think she was trying to interpret me. I thought I heard cubs in the den, so I did what the Bushmen had told me to do in such a situation, which was to walk away at an oblique angle as if I had no interest in the predator, and when I'd gone a little distance I looked around and saw this predator's hindquarters as she slowly walked back into her den. That's all that happened.

It isn't possible that she is still living, but I am, and I often think about her. I would like to have known her, not by way of a telemetric device, although such knowledge is important, but as a person, so to speak. When I had children of my own, I thought of the little sounds coming from her den. I thought of her calm but guarded demeanor when she saw me. So I shouldn't say that I envy Marcus for his intimacy with hyenas, because intimacy is the world's best way of gaining knowledge of an animal, and there's no such thing as too much knowledge about hyenas. Instead, I should acknowledge the deep gratitude I feel, and that all of us should feel, about this work that he's done and the possibilities it offers. If we knew all animals as he knows hyenas, we'd save the world.

**ELIZABETH MARSHALL THOMAS**

# ACKNOWLEDGMENTS

I am eternally thankful to my wife, Tigist, for her hard work and support during the writing process. If not for her, I would have been a very hungry writer. I am also grateful to Nigel Rothfels for his boundless enthusiasm, encouragement, advocacy, and sage advice throughout. This book is far better for Nigel's input and guidance. Thanks to Kendra Boileau at Penn State Press, not only for having faith in this project but for her encouragement, support, and critical reading of drafts at crucial times. The free rein given me by Kendra allowed me to explore some way-out ideas, some of which made it into this book. I should also thank the anonymous reviewer of my original manuscript, whose suggestions about intersubjectivity and becoming Other opened my eyes to a new perspective on my time in Harar. This provided the ideas from which I framed the book. Thanks to Elizabeth Marshall Thomas, not only for her kind words and encouragement but for unknowingly inspiring me over the years. I am also indebted to Suzanne Wolk for her copyediting. She deflected my assault on the English language and turned it into a book. Thanks to Sash Lewis for her reading of draft chapters, edits, and comments in the early stages of writing. And to Debbie Rose and Thom van Dooren, I am thankful for their guidance and input into my original thesis. I could write several pages of thanks to the many people in Harar who supported me during my time there. Some became dear friends, some gave freely of themselves, and some simply saved my sanity with a smile. Of these, I'm particularly indebted to Nureddin Adem, Yusuf Mumey, and Yusuf's family for their assistance, kindness, and love. Lastly, thanks to the hyenas of the Sofi clan, in particular Baby, Kamareya, and Willi, for admitting me into their worlds.

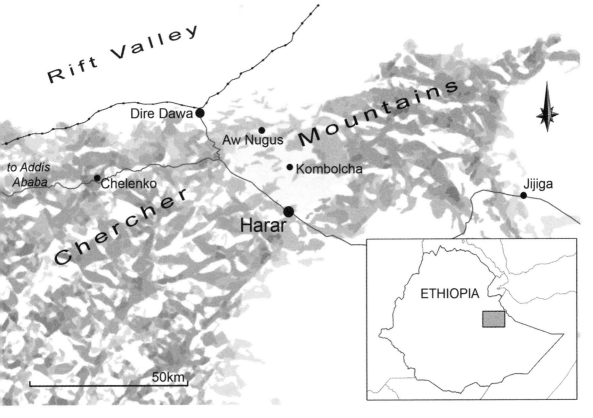

**MAP 1**   Location of Harar in the Chercher Mountains

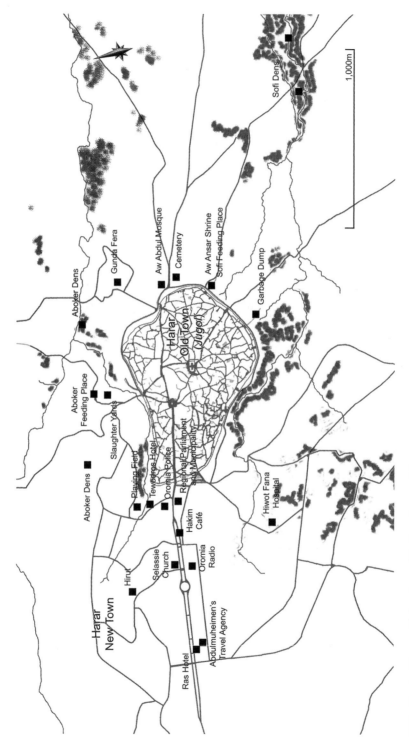

**MAP 2**  Greater Harar, with locations of feeding places and dens

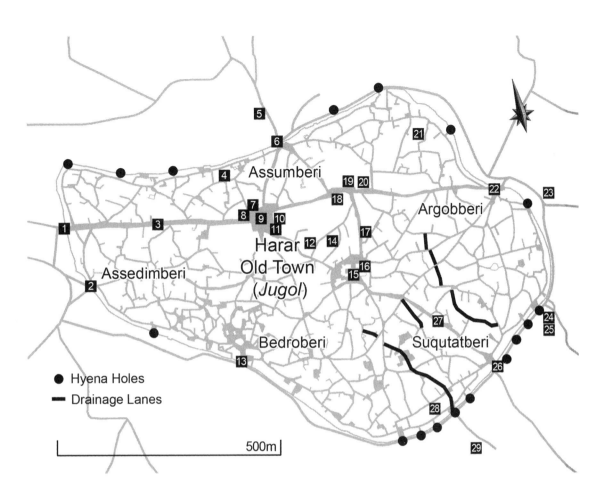

**MAP 3**   Harar's Old Town (Jugol)

1. Harar Gate
2. Gate at Assedimberi
3. Andenya Menged
4. Second residence in Jugol
5. Assumberi khat market
6. Gate at Assumberi
7. Wesen Seged Hotel
8. Emir Nur police station
9. Feres Magala
10. Orthodox church
11. Ali Bal Café
12. Abdella Sherif Museum
13. Gate at Bedroberi (Buda Ber)
14. Arthur Rimbaud Museum
15. Gidir Magala (main market)
16. Abadir police station
17. Office in Jugol
18. Tourism, Sport, and Culture Bureau
19. Jugol Hospital
20. Juma Mosque
21. First residence in Jugol
22. Gate at Argobberi
23. Aw Abdul Mosque
24. Shrine of Aw Ansar/Sofi hyena-feeding place
25. Yusuf's house
26. Gate at Suqutatberi
27. Suqutatberi Road
28. Abadir Shrine
29. Garbage dump

# INTRODUCTION

I followed the old man up a rickety staircase, taking care not to put too much weight on any steps that were split or rotten. He led me along a deteriorating hallway to room number 7. A second-floor room with a single-digit number is not unusual in Harar; the preceding number 1 had either fallen off or been unscrewed long ago, and its shadow had faded almost completely to the overall color of the door. The hotel clerk shoved it open and I dragged my bag through the doorway. I was greeted with an aesthetic slap in the face. There was a swarm of flies staging a circuit race beneath a ceiling fan that had ceased working when electricity was invented. The bed looked like it was stolen from an orphanage, after which time the original sheets had been removed in favor of a soiled set. The clerk gesticulated toward the bathroom and I went to see. I should have been more explicit at the front desk when I asked if there was a bathroom; I should have asked if there was a bathroom with running water. A small bucket sat beside the toilet, while opposite, the shower floor lay hidden beneath a layer of dried grime. It had been a long while since that shower had seen anything other than cockroaches disappearing down its drain. The clerk decided that he had no further reason to be there, so without a word he turned and walked out, leaving me to sit on the bed watching flies race around my head. Staring numbly at the stained walls, I called into question everything I had done over the past five years.

It was during the course of my honors year that I unwittingly set myself on a path to Harar. At the time, I was studying evolutionary relationships between humans and large carnivores and how these were reflected in modern humans. Because of this, I needed to find out as much as I could about large carnivore ecology, and that included reading up on hyenas. It was then that I found an intriguing passage in a seminal work on spotted hyenas by Hans Kruuk: "Only rarely are the scavenging habits of hyenas viewed with gratitude, and the most interesting example I know is the town of Harar in Ethiopia, where they walk in the streets without being molested. Occasionally they are even fed by the inhabitants and they become extremely tame."[1]

Kruuk's account of Harar describes an exception to a rule: hyenas are not well liked. In general, hyenas are loathed, vilified, feared, derided, persecuted, and, where people have the wherewithal, eradicated. Time and again I encounter negativity when I tell people about my research. They can't resist telling me how hyenas are disgusting or ugly, and they question why I'd bother studying such a hideous beast. This is a little unusual for a large carnivore; other species of this order evoke awe, admiration, and adoration, at least among people whose livelihoods are not affected by livestock predation.[2] Even wolves have overcome public relations challenges of the highest order to enjoy a mystical fondness in the eyes of those whose forebears once sought their extinction. But spotted hyenas are something altogether different: they draw the ire of Westerners as easily as that of the locals, and few ever question why this second-largest of Africa's carnivores should so excite people's prejudices. Those who do raise this question wind up scratching their heads.[3] Under the light of cultural analyses, the reasons for people's strong, almost fanatical antipathy become lost in the emulsion of comparison with other species. But there's something we're missing here. These modern attitudes are in fact germinated and grown out of the detritus of human/hyena evolutionary history. Hyenas are the old enemy. Both of our species bear the marks of millions of years of hyena predation on humans and conflict over resources. There are some unsettled scores scratched in the backs of our respective psyches.

More than four million years ago, when our arboreal ancestors thought they might do well fossicking for food on the ground, they put themselves in the path of ancestral hyenas. The hyenas soon began crunching the bones of our diminutive bipedal ancestors, favoring freshly killed specimens over desiccated carcasses.[4] Indeed, the efficiency with which hyenas demolish carcasses is a major reason why fossil evidence of the origins of our species is so scarce on the ground. Were it not for the bone-crushing capabilities of hyenas, skeletal evidence of human evolution would be everywhere.[5] Instead, much of the story of human origins and human/ hyena coexistence has been consumed, leaving only tantalizing traces of clues. Here, a collection of teeth that survived an ancient hyena's digestive juices; there, footprints in cemented volcanic dust where hominins and hyenas crossed paths.[6] What scant evidence there is of the first few million years indicates a somewhat one-sided relationship in which hyenas benefited from the presence of elusive but edible primates. But that relationship was transformed when those primates' descendants acquired a taste for the marrow of freshly killed medium-sized ungulates. At that point in prehistory, our ancestors cut in on hyenas' ancestral turf.

Enter *Homo habilis*. The appellation *Homo* confers a nonanimalness; these hominins are the first of our line to be considered and designated human. While comparatively small in stature, *Homo habilis* were well endowed with manual dexterity and tenacity; they modified stone cobbles and used the flakes and cores to process the carcasses of freshly killed ungulates over which they competed with hyenas. At least this is the origin story that comes to us from the fossil assemblages of Olduvai Gorge. While limited geographically and temporally, one assemblage of bones at Olduvai—FLK Zinjanthropus—paints an intriguing picture of an ancient coming together of meat eaters. Scratched into the bones are carnivore tooth marks overscored with cut marks from stone tools overscored with yet more carnivore tooth marks. Considerable debate surrounds the taphonomic processes evidenced by the marks, but the simplest explanation is that large carnivores killed the prey, after which hominins and hyenas competed for the lipid-rich marrow contained in the bones.[7] Needless to say, this was a direct imposition on the adaptive niche of hyenas and a direct challenge to hyenas' dominion over the dead. *Homo habilis* adopted carnivory, becoming, as David Quammen said, "more human by acting like hyenas," competing over the bodies of hoofed animals and establishing sour relations with hyenas for millions of years to come.[8]

Current evidence suggests that at the Plio-Pleistocene boundary, around 1.8 million years ago, the ranges of hyenas and hominins expanded into the Eurasian landmass.[9] This concurrent dispersal from Africa of some adaptable hominins and some giant hyenas of the genus *Pachycrocuta* was no coincidence; the two were intrinsically linked to the movements of saber-toothed cats who left large quantities of scavengeable flesh and bones for which hominin and hyaenid scavengers competed.[10] The members of this predator/scavenger guild came to occupy the entire Eurasian landmass, from modern-day Ireland to Mongolia, from Germany to Java.[11] In combination with the already established spotted hyenas, these meat and bone eaters established a continuous human/hyena presence that persisted through dynamic climate changes and variable faunal compositions until the end of the Pleistocene ten thousand years ago. By that time, modern humans had even made art of spotted hyenas in ochre and antler.[12] This is not to say that humans tolerated their subjects; the appearance of modern humans in Europe coincided with the rapid disappearance of hyenas. Mary Stiner suggests that wolves outcompeted hyenas in newly forested environments, but let us not discount the capacity of the bipedal ape to shove other species aside.[13] Just prior to ten thousand years ago, spotted hyenas disappeared from Europe and Asia. A dwarfed variant emerged in the

Middle East, perhaps representing a last effort to adapt and persist, before spotted hyenas completely disappeared from the lands beyond the African continent.[14]

Under the shadow of millions of years of conflict, is it any wonder that relations between humans and hyenas in Africa are generally strained? No wonder, considering how our two species have such large range requirements and the same fondness for cows, goats, and sheep. Competition between humans and hyenas over livestock resulted in the complete eradication of spotted hyenas from southern Africa, their extinction in several other countries, and the contraction of their range all over the continent.[15] Despite numbering twenty to forty thousand in Africa, current circumstances and the general trend indicate that hyenas are on their way out.

The way that hyenas comport themselves doesn't help their cause. Their times of greatest activity, vocalizations, physical attributes, and ecologies combine with people's fears and prejudices to produce some extremely negative conceptions. Often, hyenas are associated with what are called witches.[16] Hyenas' nocturnal habits, humanlike giggles, and propensity to exhume and consume human corpses make them obvious companions to people who use supernatural means to malevolent ends. Often, the hyenas are believed to do the bidding of witches, bringing home fresh human meat, or else they are ridden through the night sky with flames shooting out of their anuses. In some places, witches transform themselves into hyenas and prowl around at night looking for victims to eat.[17] In this respect, killing a hyena is not always about pest control. It is a way of killing or disempowering malevolent humans.

Beyond the association with witches, hyenas' necrophagy is dangerous in other ways. Where people dispose of the dead by leaving them out for hyenas to consume, a hyena's defecating in the village—returning the dead to the living—can be a serious concern.[18] Hyenas also prey on the living. They attack people sleeping outdoors in the hot summer months, or they break into huts and tents and drag their victims away to be eaten.[19] They are especially adept at preying on children, and unfortunately for the victims, the usual method of attack is to bite the face. Even when children survive, they bear the scars of the attack for the rest of their lives; they are walking reminders of the dangerous creatures lurking outside the village.

Hyenas are also loathed and derided for their confusing sexuality; the enlarged genitalia of the females lead people to see them as hermaphrodites. And hyenas are a joke, a hated joke. African folktales consistently portray them as loathsome, greedy, and above all stupid.[20] No doubt their appearance influences such conceptions. Ungainly heads, perched on the ends of overly long necks, bob up and down as they lope off into the long grass; falsetto voices call out to nowhere in particular, while hard feet designed for long-distance running slip out from under their bodies on wet surfaces. During a safari in Kenya, Ernest Hemingway shot a hyena who tumbled to the ground and tore at her wound. He gave an account of the reaction of his Kenyan guide:

> "*Fisi*," M'Cola would say and shake his head in delighted sorrow at there being such an awful beast. Fisi, the hyena, hermaphroditic, self-eating devourer of the dead, trailer of calving cows, ham-stringer, potential bit-er-off of your face at night while you slept, sad yowler, camp-follower, stinking, foul, with jaws that crack the bones the lion leaves, belly drag-ging, loping away on the brown plain, looking back, mongrel dog-smart in the face; whack from the little Mannlicher and then the horrid circle starting. "*Fisi*," M'Cola laughed, ashamed of him, shaking his bald black head. "*Fisi*. Eats himself. *Fisi*."[21]

And yet there remained Kruuk's account of Harar: a town where hyenas walked the streets at night and were "encouraged by the local population," a town where men sat outside the town wall at night, handing scraps of meat to hyenas.[22] It was an inconspicuous passage, but it stuck in my mind. When the time came for me to plan my doctoral research, I was interested in large carnivores and anti-predator adaptations in humans. While I thought I should at some stage go to Harar and collect some limited data on attitudes toward hyenas, it was my honors coordinator, Marcus Barber, who thought this was worth a bit more of my atten-tion. He convinced me to write a research proposal for a dedicated ethnographic study of hyenas and humans in Harar. At the same time, Macquarie University's Centre for Research on Social Inclusion made a scholarship available for doctoral research into human/animal relations. My proposal fit nicely with what they were looking for, so under the supervision of Debbie Rose, I began making plans to spend a year in Harar doing ethnographic research on people and hyenas. It's funny the places life takes you.

As far as ethnographic research goes, Harar is a pretty good choice of field sites. Just north of the equator and 1,850 meters above sea level, the weather is near perfect. There is power, phone, internet, and banking to make things easier for the researcher. But over and above that, the Harari people are fascinating. They claim that Harar is the fourth-holiest city in Islam, after Mecca, Medina, and Jerusalem. Supporting their claim, eighty-eight mosques and 121 Sufi shrines are located within the half square kilometer that is Harar's Old Town. The Harari people retain their own language, clothing style, architecture, and traditions amid a large population of settlers and settler descendants from among the eighty other ethnic groups in Ethiopia. In the Harari language, the town is known as Gey, meaning "the City," and nearly everything associated with Harar is prefixed with *gey* to denote that it is "of the city." Hararis are *gey usu*, the Harari language is *gey sinan*, local baskets are *gey mot*, houses are *gey gar*, and indeed the local hyenas are *gey waraba*. These last seem incongruous, though, in a modernized city with a population of a hundred thousand and a history dating back more than a thousand years.

In the accounts of Harar by various writers since the nineteenth century, the hyenas often lurk in the shadows, carnivorous functionaries appearing here and there but never central to the texts. Both Richard Burton and Philipp Paulitschke, who visited Harar in the nineteenth century, mentioned the presence of hyenas. Burton's mention sits within an account that reads as a precolonial reconnaissance mission on behalf of an expanding empire. Alongside his description of Harar's inadequate defenses and of a populace needing a colonial power to step in and save them from themselves, Burton wrote that the hyenas in Harar were "lured" into the town, after which the gates were closed behind them and they were "safely speared," a pest easily eradicated according to the fashion of the era.[23] Paulitschke related his own experience of hunting hyenas in the hills around Harar. He was told that in times of famine, hyenas entered into the town and attacked people in the streets. Paulitschke also described how emirs in former times organized hunts, not to kill hyenas but to "worry" or "dislodge" (*beunruhigen*) them from their hiding places.[24]

Later, in the 1960s, Wolf Leslau compiled and translated some accounts of Harar written by Harari men attending the University of Addis Ababa. In an account of Harar's hyenas, one of Leslau's informants relates that the hyenas emerged from their dens at night and ruled the land outside the town's defensive wall.[25] Whether they entered the town at night is not clear. The compilation also includes an account of the festival of Ashura and the tradition of feeding porridge to hyenas during the celebrations.

The most recent mention of hyenas in an ethnographic account of Harar comes from Camilla Gibb, who conducted fieldwork there in 1994. Gibb wrote that Hararis regarded the hyena as a "curious sort of half brother, half wild, half civilized, roaming the hills around the city yet familiar with the tangled streets of the town."[26] She also wrote about the practice of feeding hyenas in Harar. There was a man who entertained paying tourists by feeding scraps of food to hyenas at a place just outside the town wall. According to Gibb, the practice was sold as a tradition but was really only a money-making venture. If there were no tourists at the feeding place, the hyenas were not fed.

I didn't realize it at first, but my decision to include hyenas in an ethnography made for some unexpected theoretical and methodological challenges. Anthropology by definition is the study of humans, and anthropology's tool kit—ethnography—is designed for recording and interpreting the beliefs and behaviors of humans—and only humans. Traditionally, animals feature in ethnographies as things that people hunt, eat, ride, think about, or use as symbols; they are almost never considered as participating members of human societies.

In the early 1990s, the Dutch anthropologist Barbara Noske criticized this state of affairs. She argued that representations of animals in anthropology were drawn from assumptions based on biobehavioral accounts rather than firsthand empirical observations. Noske called for an anthropological approach to the study of animals. She argued that an "inter-subjective non-reductionist" approach was necessary before social scientists could decide what animals were or were not. Furthermore, the field of anthropology, with its intersubjective methods of understanding, was well situated to adopt such an approach.

After deciding on balance to include hyenas as members of the society that I was studying, I immediately encountered difficulties. It turned out that I had to apply for two sets of ethics approval and get two sets of permission—one for human research and one for wildlife research. I also had to come up with some research questions that pertained to relations between humans and hyenas. I struggled with that for some considerable time before letting it go. I was faced with the difficult task of collecting data from two different species and bringing those data together in a dissertation with some kind of theoretical underpinning, when there was no established theory other than Noske's call to "bring in the animals." I also needed some kind of methodology. Normally, biologists follow hyenas around in vehicles making observations, but even if I could have included a four-wheel-drive vehicle in my budget, it would have been useless. Of the hundreds of lanes and roads in Harar's Old Town, only five are wide enough to allow the passage of vehicles. In the farmland outside Harar, gullies, woods, streams, and thorn hedges break up the landscape, making it traversable only on foot.[27] While I was struggling with methodology, I was given a timely copy of Elizabeth Marshall Thomas's book *The Hidden Life of Dogs*.[28] In order to gain insight into the nocturnal movements and activities of a dog named Misha, Marshall Thomas simply followed her subject around the neighborhood with a notebook. This sounded like it could work. I could spend a lot of time at the place where hyenas were being fed for tourists and habituate them to the point where they'd let me follow them around Harar recording my observations.

To that end I packed a bag full of technology for making nocturnal observations, recording interviews, and taking notes on the fly. I brought a handheld voice recorder, a flashlight, a camcorder with night-vision capability, an infrared searchlight, a night-vision monocular, a laptop with external memory, a tripod, and an

SLR camera with three lenses suited to varying degrees of closeness.[29] I think that if the customs officers at Addis Ababa airport had bothered to inspect my bags, they would have arrested me for spying. I had neither intention nor sufficient training to use tranquilizer darts and radio collars. While I can understand how radio collars and receivers make it extremely easy to locate hyenas, and how GPS devices provide very revealing data that would otherwise be almost impossible to obtain, I have reservations about imposing a device on an animal who would rather not have to wear it. There were also cultural issues I needed to consider; in light of what I had read in other ethnographies, putting a collar around the neck of a hyena might have been offensive to some people in Harar.[30]

So it was that with a suitcase full of equipment and very little else, I stepped out of Dire Dawa airport into the dry heat of the Rift Valley. After the temperate vegetation in Addis, the acacia trees surrounding Dire Dawa appear almost as an African cliché. Whereas cold and rainy Addis seems anomalous, you really feel like you are in the stereotypical Africa when you arrive in Dire Dawa. The airport terminal is a sparse building in the style of the 1960s, complemented by the old Peugeot taxis waiting in the car park for the infrequent flights to arrive. I shared a taxi with a local doctor who was being stationed in a rural clinic, and went to a hotel where I left my bags. Before I traveled on to Harar, I had to visit the immigration department and apply for a residence permit, so I rode a three-wheeled "Bajaj" taxi through the tree-lined streets to the first of many frustrating and inevitably fruitless encounters with Ethiopian bureaucracy. After being told that I could not have a permit but could "simply" return every month and have my business visa renewed, I collected my bags and went to the bus station, from which I could get a minibus to Harar.

The minibus from Dire Dawa to Harar ascends a mountain range and crosses a verdant plateau to its eastern edge, whereupon Harar sits on a rocky spur. In my ignorance of what was standard practice, I protested that the minibus I was on was overcrowded and insisted on being let off to take another. My protest was greeted with much hilarity as the ticket collector slid the door shut and we were off. Consequently, I glimpsed only parts of the journey through the spaces between the bodies of my fellow sardines. I saw wattle-and-daub huts lining the roads, alongside which little earth-dusted children played; men whacked oxen with sticks as they carved furrows with wooden ploughs roped to the necks of the beasts; women balanced handwoven baskets on their heads as they followed their donkeys to market. It was like a journey into the Middle Ages on a minibus, but with modernity tacked on in all sorts of places. The women talked into mobile phones as they marched along; some of the wattle-and-daub huts had shiny tin roofs, and the children played with the lids of Coca-Cola bottles.

Arriving in Harar, I was ejected from the crowded minibus like a cork. I immediately looked to the roof of the vehicle and was amazed to see my bag where I expected to see only a dangling rope. It took three of us to drag the bag down from the roof successfully, and I wondered how much of the technological contents had survived the journey. I was already beginning to draw the attention of the locals, and I'd only been on the ground a few seconds. Some children prodded me and a leper waved the stump of his wrist in my face. I pretended to ignore them and hailed a taxi, asking the driver of the old Peugeot to take me to the Tewodros Hotel. I had read good things about this hotel. It was close to the Old Town, it had bathrooms, and it was relatively cheap. I was optimistic, even when I found reception a dingy office with a table and little else. I waited a few minutes in the office until a dusty-looking clerk emerged from the bar and went solemnly behind his desk. When I inquired about a room, he regarded me with a mixture of suspicion and pity. He shrugged his shoulders, told me the price, and led me upstairs to my room of flies. As I said, it's funny the places life takes you.

# I

## PAST FINDING AROUND HARAR

When I eventually emerged from my room, the clerk was waiting for me down-stairs. He waved a hand toward the end of the landing under the veranda and said, "This is your guide." The only person I could see was a man in a stained T-shirt and dusty jeans lying on the floor in front of room number something. As the man looked up at me, his eyes widened. I didn't realize it at the time, but even on my meager university stipend, the asymmetries in per capita income between my own country and Ethiopia had rendered me among the wealthiest persons in Harar. What's more, my white skin was like a close-fitting banner, advertising to one and all that not only did I have a lot of money but I was naïve about the town and the local rates of pay. I was therefore a walking, talking opportunity to anyone who could speak sufficient English and was able to convince me that they were necessary to my survival.

The man dropped the khat leaves he was chewing (pronounced *chat* in Ethiopia), stood up, wiped his hands on his trousers, and introduced himself as Nassir. As we shook hands, I noticed the brown spots that mottled his teeth and eyes—the marks of the khat-afflicted. Nassir's round face and chocolate skin hinted that he was from someplace other than Harar. It turned out that he was Gurage but had spent enough time in Harar to have learned the local languages and established himself as a tour guide. He asked me what I wanted to see in Harar: museums, the Old Town, shrines, hyenas? I told him that I was there as a researcher, intending to spend a year in the town—at which his eyes opened wider—and that I really only needed to find rental accommodations. That, and to make arrangements to do research at the place where the man fed hyenas for tourists. He told me no problem and gestured for me to take a seat while he made several phone calls, alternating between different languages as he spoke to different people. After he was satisfied that we had enough options to explore, we left the hotel compound and flagged down a Bajaj taxi to take us into the Old Town.

The three-wheeled taxis in Harar are different from those in other parts of Ethiopia. Whereas those in Dire Dawa and Adama are relatively quiet, petrol-powered models, the taxis in Harar have thumping diesel engines that rattle out a raucous "tuk-tuk-tuk!" as they ply the streets of the town.[1] Nassir explained that the noisy diesel models were prohibited in other Ethiopian towns, but that in Harar, anything goes. Moreover, they had only been around for the past six years; prior to that, the streets were peaceful. The old 1960s Peugeot taxis and horse-drawn two-wheeled carts—I hesitate to call them hansom cabs—were steadily diminishing in number as Bajaj taxis predominated. Our Bajaj passed through Harar gate and down a crowded and comparatively wide street toward the center of Harar's Old Town. The driver swerved around a deformed man who was hunched in the middle of the road. His legs were bent backward over his shoulders, so he propped himself up and shuffled about on sandal-shod hands. We rounded a bend and jerked to a halt in front of the crowds of people outside Jugol Hospital. A leper sat outside the hospital gate in a ramshackle wheelchair, holding out the stubs of his fingers and making plaintive cries as people jostled around him. The people carried on, oblivious to his pleas, as they vied for seats in the taxis. Only the poor, sick, and suffering seemed to have any fixity; the remainder of the population hustled around them in a blur of activity.

The name Jugol applies to three different but related entities in Harar: the wall, the entire Old Town, and a central place within the Old Town. Jugol the wall is a stone-and-mud construction averaging four meters high and punctuated by six main gates as well as several smaller entrances. At its western end, the wall is smooth and capped with triangular battlements. Elsewhere, the stones and mortar are visible and knobby. Inside this wall is Jugol the town, a little more than half a square kilometer of densely packed stone buildings that share compounds surrounded by three-meter stone walls. These walls within a wall line the sides of more than 350 lanes, constituting what is best described as a maze. Just five roads in Harar's Old Town are navigable by vehicles. The remainder are traversable only on foot, or in some cases, and much to the chagrin of pedestrians, by motorcycle. There is also a central point in the Old Town known as Jugol. This is where taxis turn around in front of an old stucco building with a small forecourt: Jugol Hospital. Hence if you are coming from the New Town and you want to pass through Jugol the wall and go to the southeastern part of Jugol the town, you ask the driver to take you to Jugol the place in front of the hospital.

Nassir led me down a lane opposite Juma Mosque, the preeminent mosque in Jugol the town. The walls of the lane rose up on either side of us, so it became difficult to discern in which direction we were traveling. This was something with which I was to become familiar: navigating by landmarks rather than by an overall picture of the landscape in my head. We arrived at a gate, and after Nassir knocked continuously for a minute, a very old, very small Harari woman let us into the compound. A couple of Harari houses shared the compound, and above one of them, at the top of a precarious flight of stairs, we were shown one of two single rooms. It was small; it had an adjacent toilet down which you could pour water and a shower floor for washing with a sponge. Nassir spoke for a while with the old woman and they arrived at a price. The cost was comparable to that for a room in my own country, so I declined and we left. After visiting another house with a room that I was assured was excellent but was not allowed to see, we took a Bajaj out into the suburbs of the New Town to look at a relatively new but overpriced house. I decided to hold out until I found a better option, so I arranged to meet up with Nassir that evening to visit the hyenas and rode a Bajaj by myself back into Jugol to do some exploring.

I was interested to see the shrine to a man named Abadir. Of all Harar's historical figures, Abadir is by far the most prominent, and at the same time one of the most enigmatic and elusive. His origins are uncertain, but it's most likely that he came from across the Red Sea in what is now Yemen. About a thousand years ago, Abadir journeyed to Harar in the company of forty-three male and female Sufis, and they set about teaching Islam in the region.[2] The Sufis established schools in various locations around Harar, while Abadir stayed in the town and eventually became somewhat influential. He established many of the town's social institutions, became Harar's second emir, and defeated Harar's enemies.[3] There is no record of who Harar's enemies were at the time—possibly they were Abadir's predecessor and his supporters, as the townspeople have occasionally had strained relations with their leaders—but there is no doubting Abadir's influence. Weekly celebrations are held at Abadir's shrine, where religious songs are sung to celebrate his personage.[4]

Walking down one of Harar's five vehicle-friendly roads to the main market, I was followed by three small boys calling out, "*Farenjo*, money!" Farenjo is the local term for a fair-skinned foreigner.[5] This must have been a variation on *farenji*, which is common in other parts of Ethiopia and probably a loanword from the Arabic *faranj*, which historically referred to a Frank. By the time I'd entered into a narrow lane beside the entrance to the main market, I had "farenjo" ringing in my ears. I heard it from children, from behind shop counters, from market stalls, and from old men who passed me in the street. A skeletal woman sitting with her back against a wall held out a desiccated hand and in a pathetic voice moaned, "Farenjo, baksheesh." I had money. I wanted to give her at least one Ethiopian birr for a couple of bread rolls, but I had a gaggle of children at my heels, and I knew I could not easily do a year's research on hyenas in Harar if I became known as a walking money dispenser. Instead, I smiled an embarrassed smile at the woman and walked past, adding a little layer of hardness to my heart as I headed down-hill through a maze of lanes. After briefly emerging from the maze at one of the metaled roads (a metaled road has a stone aggregate and bitumen surface) and regaining my bearings, I followed another lane to a large walled compound in the shadow of two giant fig trees: Abadir.

At the entrance to the compound, there was a scarlet-colored gate framed by a whitewashed arch with bright green trim and a crescent moon and star be-neath the arch. I peered in, looking for a turquoise beehive-shaped tomb, typical of Harari shrines, but all I could see was a complex of buildings on either side of a narrow lane. In Harari terms, it was opulent. The whitewash gleamed, bright green paint trimmed the walls and roofs, and floodlights mounted here and there suggested nightly celebrations. It spoke of the importance of this man, Abadir, and his influence continuing on through the centuries. But the tomb itself appar-ently lay hidden inside the complex of buildings, like his personage beneath so many centuries of embellishing legends. I declined to go in. Not having brought a gift of khat or incense, it was inappropriate for me to pay a visit, so I turned and headed along a lane that would take me to Jugol the wall. I wanted to see a *waraba nudul*: a "hyena hole."

I'd read about these hyena holes in Camilla Gibb's dissertation. Imagine: holes built into a wall to admit hyenas. It was a kind of physical expression of the human/hyena relations in Harar by which I was fascinated and about which I needed to understand more. But Camilla Gibb had also included a disclaimer: there was de-bate among Harari historians over whether the holes had been built into the walls to admit hyenas or whether they were simply for drainage. Moreover, they could have been intended to serve a dual purpose. Regardless, I was drawn to them, if only for their names. I rounded a bend in the lane I was following and there before me was a meter-high hole in the wall.

I scanned the ground looking for hyena footprints, as I still had no idea whether hyenas were coming into the town at night to look for food. Nassir had assured me that they were, but he had assured me of a lot of things, such as the ease with which I could find accommodation in Jugol, so I was looking instead for some physical evidence. The problem was that the ground on either side of the hole was covered in refuse, which had piled up as the rains tried to rinse it from the town. To my untrained eye, it was impossible to distinguish any kinds of prints among the plastic bags, leaf litter, and indistinguishable brown matter. And in front of the refuse the ground was hard, gravelly, and disturbed by countless human feet that had walked past the hole. It certainly looked large enough to admit a hyena. The small lane leading to the hole provided sufficient cover for a hyena entering the town to wait for the coast to become clear. But my doubts lingered as I left the hole, looking for an exit through the wall that could take me to the world outside Jugol.

The exit I found was another hyena hole, but this one high enough and sufficiently clear of refuse to allow the passage of people. As I approached, a lean man passed through the hole with his head bowed so as not to hit the wooden beam across the top. The man was Oromo. He wore a colorful cotton skirt and a striped polo shirt, both of which were almost a uniform for these people in the rural areas outside Harar. He was also carrying a bag of khat leaves and a long-handled sickle, known as a *muncha*, which rested on his shoulder. The man lifted his head after passing through the hyena hole and noticed me. He looked at me with a mixture of curiosity and suspicion as he walked past and disappeared into the maze of lanes. I found some irony in watching an Oromo man enter Harar via a wall that had been built for the express purpose of keeping his people out.

In the sixteenth century, Harar was the capital of a vast sultanate under the leadership of a charismatic man known as Ahmed Gragn, the "left-handed." Gragn himself was not a Harari—most writers suggest he was Somali—but he made Harar his capital after marrying a Harari woman, Bati del Wambara. From his base in Harar, Gragn waged jihad against the Abyssinians in the west, destroying churches and forcing conversion upon the vanquished. After Gragn's death at the battle of Wayna Daga, however, another player emerged: the Oromo. The historian Donald Levine argues that years of war between Muslims and Christians created a political and military vacuum that the Oromo rapidly filled, albeit incidentally to the practices associated with Oromo social institutions. Without a centralized government, and with a system of age sets that required that each community wage war against a neighbor every eight years, the Oromo soon reached critical mass.[6] The boundaries of their territories expanded organically and rapidly; they simply absorbed other peoples as they "broke like a tidal wave over Ethiopia."[7] The wave swept on toward the Chercher Mountains and Harar.

Back in Harar, Bati del Wambara married Gragn's nephew, Nur b. Mujahid, apparently on the condition that he avenge her first husband's death.[8] Despite fulfilling the condition and continuing the holy war against the Christians, Nur's challenges were compounded by the increasing strength of the Oromo. When an Abyssinian army sacked Harar while Nur was away campaigning against another, Nur hurried to rescue the town, but on his way he was engaged in battle by an

Oromo army near Awash. While both sides suffered heavy losses, Nur was in the end victorious and returned to Harar to find the city devastated. Realizing Harar's vulnerability in the face of Abyssinian attacks and the Oromo expansion, Nur ordered the construction of a defensive wall around the town. Over the following centuries, this wall not only preserved the Hararis from outside foes; it defined the populace as a people at odds with the outside world. The Oromo did come to occupy the landscape all around Harar, but they never took the town. The once vast sultanate became a little Harari island in a vast Oromo sea.

As I passed through the hyena hole with my head bowed, I was greeted by a cool breeze and a sweeping view of the outside world. To my right, Mount Hakim towered above the ramshackle houses on its lower slopes, dominating Harar's southern aspect. In front of me, the land sloped gradually downward toward the arid lands of the Ogaden, where the streams from around Harar became rivers and then eventually dried up without ever reaching the Indian Ocean; to my left, the Chercher Mountains regained their heights, culminating in the distinctive W-shaped profile of Mount Kundudo far off in the distance. I turned and contemplated the wall. It was unimpressive as a defense: patchy in places, presumably from restoration work, barely high enough to impede a determined climber, and definitely knobby enough to offer plenty of purchase. I wondered at the Oromo people's adherence to fighting on horseback and how much this might have contributed to Harar's remaining intact.

I headed north along the dirt road that ran alongside the wall for most of its length. Some grubby children playing outside patchwork huts on the opposite side of the road called out "farenjo," giggling uncontrollably at their own temerity. A woman put a hand up to balance the basket on her head while she squatted over the drain that ran between the wall and the road. Her striped polo shirt indicated her ethnicity, while her long skirts disguised whatever bodily function was engaging her. A small boy drove an ox past me, whacking the beast repeatedly with a stick. The animal, twenty times the boy's weight, showed no inclination to slow down or stray from the road. The boy looked at me and said "farenjo." I was passing a gate named Suqutatberi. The name comes from *sugud attay*, meaning "place where there is a pile of bones," and it probably reflects the gate's location at the lowermost point of Harar. I imagined that anyone in Harar with a load of bones to dispose of would be inclined to take it there rather than uphill to another of Harar's gates. I also imagined that Suqutatberi's amenity for the dumping of refuse must have made it an attractive place for hyenas since the town's inception. Little wonder, then, that it was near this gate that hyenas were being fed on a nightly basis.

I had read about Harar's hyena feeders in Camilla Gibb's thesis and in the Brandt travel guide. During the time of Gibb's fieldwork there was only one hyena man, but by the time Philip Briggs made his entry in the travel guide, there were two. I had also seen video and photos of these hyena men on the internet. One of these men, Yusuf Mumey, fed hyenas just north of Suqutatberi, next to the shrine of Aw Ansar. As I approached the place, I was struck by the strange appearance of the shrine. It was a whitewashed beehive tomb with a green roof and green trim, typical of Harari shrines, but it had an unusual feature: a giant fig tree growing through it. Actually, it was difficult to discern which was there first, the tree or the tomb, as each seemed to have incorporated the other in its structure. I stopped on the road and looked past the shrine to the open area in front of it. A couple of chickens were scratching around in the dirt. There were no hyenas around, as it was midafternoon, but plenty of half-chewed bones were scattered beside the road. I wondered how this place must appear in the dark of night, crowded with people and hyenas.

Walking uphill, I found myself between two walls. That on my left was Jugol, that on my right was the boundary of Harar's Muslim cemetery. As with the town's refuse, it was easier to carry the dead downhill than up. The Hararis had a system of neighborhood groups known as *afocha* to which members contributed money. These afocha were responsible for the costs and conduct of members' weddings and funerals, so they ensured that deceased members were buried in a niche at the bottom of a deep grave. The homeless, by contrast, were the responsibility of the municipality. They were unceremoniously deposited in shallow graves, easily accessible to grave-robbing hyenas. There was plenty of opportunity at the downhill end of a densely populated town for a carrion-eating carnivore. As I continued on, a boy walking down the hill looked at me, smiled broadly, and said, "Farenjo, fuck you, okay!"

At the top of the hill, I passed another gate named Argobberi, so named after the Argobba people, who constitute a minority in the region. This ethnic group is enigmatic, both linguistically and geographically. Divided into two populations, one on either side of the Rift Valley, they are generally Muslims, yet their language closely resembles that of the Christian Amhara. The Argobba east of the Rift, in the Harar region, have apparently lost their native language and speak only Oromo, and in one place Harari.[9] Moreover, their settlements lie separated and scattered across the region on hilltops, just above the Oromo high-tide mark. Outwardly, the Argobba resemble the Oromo in their clothing, although the beadwork that the girls wear can be quite ornate. In correspondence with a Harari living in Australia, I'd raised the subject of the Harari children's chant, *Argobba: Maltu maltu, usu; mishet mishet, waraba* (Argobba: by day, people; by night, hyenas). He dismissed it as meaningless and insisted that it reflected neither Harari attitudes toward the Argobba nor any theriomorphic capacity on the part of the people themselves.

Passing the outer suburb of Gunda Fera, I followed the encircling road west-ward, where I crested a hill and could see some of Harar's New Town. An Oromo man wearing a *coloyta*—a Muslim cap—stared at me as he approached and said, "Farenjo." The majority of the Oromo population in the Harar region is Muslim, but this was not always the case. Prior to their great, organic expansion, the Oromo adhered to a monotheist/animist belief system in which a remote sky god was manifested by spirits that inhered in animals, trees, rocks, and even imple-ments. This religion apparently prevailed until the Egyptians occupied Harar and the surrounding region in the 1870s and '80s. While in occupation, the Egyptians set about defeating the Oromo chiefs and enforcing conversion upon the people. This increased the influence of shrines in the region as places of teaching Islam. Still, conversion was an ongoing process, and the national census of 2007 found a few thousand Oromo in the region who still adhered to the old beliefs.[10] As for the Egyptians, they departed Harar in 1885, after building a few fortifications, recon-structing a gate (Assedimberi), and installing a Harari emir named Abdullahi.

I arrived at Assumberi, the gate for which the name is not as determinate as the others. The road I was on ended at a T-junction that offered a choice of going left back into Jugol or right toward the Oromo town of Falana. On both sides of the road lay the khat market, where Oromo women sat under umbrellas and tar-paulins with bags of the emerald-green leaves while men crowded around, testing the quality, checking for signs of pesticides, haggling over price. Fifty years earlier, it was Harari women who sold khat at the market; they bought the leaves whole-sale from Oromo farmers and sold them at a profit.[11] But the Oromo were increas-ing their influence in the region; the women had wrested control of the khat trade from the Hararis, and the gold jewelry adorning their fingers, wrists, ears, and noses attested to the profitability of their enterprise. Meanwhile, Harari men (and a few women) were buying into this power shift, daily handing over considerable sums of money for the bags of green leaves so important to social, religious, and business life in the city. Another Oromo tide—but this one green and flowing into Harar daily, while the money of the city daily ebbed.

I knew that another hyena-feeding place lay beyond the khat market, on the northbound road beside the slaughter yards. Again the emergence of hyena feed-ing at a place where hyenas already had reason to congregate. But I decided I would not learn much by visiting at that time of day, so instead I turned and headed back into Jugol through the opening in the wall where once there was a wooden gate and stone arch. A man passing through on his way down the hill to the khat market said, "Farenjo, welcome," as I began the ascent of the hill toward Jugol Hospital along a steep metaled road speckled with animal droppings. A woman was walking down the hill with several goats who were apparently oblivious to their impending fate at the slaughter yards some three hundred meters farther on. Another woman was leading a donkey up the hill, the beast exerting itself under a burden of sacks of flour and a wooden pack apparatus that weighed almost as much as the load. My breathing became labored as I continued up the hill.

During the reign of Emir Abdullahi, the colonial gaze was fixed on Harar and the Hararge region. Hence the emir, not without justification and possibly augmented by khat, became somewhat paranoid. He ordered all non-Hararis in the town to either convert to Islam or else get out.[12] When he received news of an Italian trade reconnaissance mission on its way to Harar, he ordered a squad of troops to intercept the mission and kill the Italians. Then he changed his mind and sent another squad to intercept the first and rescind the order. All too late, as the Italians were massacred. But Abdullahi's paranoia found justification when news came of an Abyssinian army under Menelik II marching across the Rift Valley and into the Hararge region with the intention of taking Harar. With all haste, the emir raised an army of Hararis, Oromo, and Somalis, armed them as best he could, and marched with them to a place called Chelenko. There, they occupied the heights, fortified a hilltop, and waited for the arrival of Menelik's army.

The place, Chelenko, is unremarkable, a narrow plain with a patchwork of khat farms bounded by low, rocky hills. But the name Chelenko is one of the most evocative in the history of Harar. It speaks of a place of utter defeat, where Harar finally lost its centuries-long struggle to maintain autonomy. Menelik arrived at Chelenko on January 5, 1887, with twenty to twenty-five thousand troops, about half of whom were armed with modern breech-loading rifles. His army made camp on the plain below the heights, where the Harari army looked down from their advantageous position at their tremendous numerical disadvantage. They numbered three to four thousand poorly equipped troops who were more adept at hand-to-hand fighting than at gun battles. Abdullahi had originally intended to defend the heights, but the timing of the Abyssinians' arrival presented him with what he perceived as an opportunity. The day after Menelik's arrival, January 6, was Ethiopian Christmas, and the Abyssinian troops were expected to be occupied with celebrations for the day. Abdullahi decided to launch a surprise attack, but Menelik was wise to the ploy and had his army ready. When the Muslims came swarming down the hillside, expecting to fall upon an undefended camp, they ran into a hail of bullets and cannon fire. Meanwhile, Menelik sent his cavalry to the heights, where they overran the Harari cannon position. Abdullahi, on seeing his army defeated, fled for his life, bypassing Harar and escaping to Jijiga in Somaliland.[13] He hoped to continue the holy war from there but played no further part in Harar's history, eventually dying in obscurity. The survivors of his army fled for their lives, while Menelik marched on into Harar and installed a governor. Menelik ordered a gate built in the western side of the wall, a wide road cut through the town leading to the central square, and a mosque demolished, to be replaced by a church at the edge of the square. All things considered, he was quite lenient. He left the Hararis with a considerable degree of self-determination that was only later reduced by succeeding governors.

In Harar, Chelenko and the loss of independence are symbolized by the color red. Traditional Harari houses have five seating platforms in the living room, the number apparently representing the five pillars of Islam. The bases of these plat-forms are invariably painted red, reminding anyone who enters the house of the blood spilled by so many Harari men at the battle of Chelenko. When a Harari man marries, he wears a folded scarf with a narrow red stripe over his shoulder. The scarf, known as *satti baqla* (seven hundred), is an overt reminder: seven hun-dred Harari men were killed at Chelenko, not an inconsiderable number given Richard Burton's estimate of a total Harari population of eight thousand in the 1850s. The red stripe in the state flag and the red in the logo of the Harari ruling party also represent Chelenko. When you become aware of the symbolism of the color red, you begin to see Chelenko everywhere—in the red squares of Harari basket patterns and the red detail of the traditional Harari headwear for men, in all kinds of red-colored things.

Arriving at the top of the hill, I turned right and headed toward the central square, known as Feres Magala, or Horse Market. There were no horses to be seen, only Bajaj and old Peugeot taxis turning in front of the church gates. Christian Amhara women, their heads covered in white scarves, jostled for places in the taxis headed out of the Old Town. If not for the church that Menelik built, they would have had no cause to be in Jugol. Most Amhara found the maze of lanes intimidating and rarely ventured into the Old Town. Instead, they lived their lives in the New Town, established after colonization, never knowing the narrow lanes behind the wall.[14] On church days, they took taxis into Jugol, observed the rites at the church, and then took taxis straight back out again. As I passed, one Amhara woman tried to squeeze into an already full Bajaj and was unceremoniously ex-pelled as it took off up the road. Demolishing a mosque and putting a church in the middle of a Muslim town is not just disturbing for the Muslims.

Across the square from the church, I could see the road that Menelik built leading all the way to the gate in the wall. He named the road Andenya Menged, or First Street, as though the existing streets were inconsequential. In comparison to the other roads and lanes in Jugol, this one was very wide. Three vehicles could probably drive abreast along its length and still leave room for a few pedestrians. Looking at the chaotic traffic, I guessed three vehicles probably already had. On either side, little shops sold hardware, jewelry, textiles, and stationery. There was also a mosque halfway along, and a couple of hotels. I noticed in one of the hardware shops an old Oromo man wearing a traditional cotton skirt and holding a muncha over his shoulder. The man was buying a generator. Meanwhile, Oromo women marched past me with baskets of goods on their heads and donkeys walking ahead of them. One of the women glanced sideways at me but her expression remained the same—determined—and she continued on her way. The others in her group, refreshingly, ignored me completely.

At the top of Andenya Menged, I passed under the decorative gate that Menelik built, known as Harar Ber (Harar Gate). This was a wide central gate for vehicles, with smaller arches on either side under which pedestrians could pass. The battlements on top were punctuated by inverted semicircles, a style that had since been adopted in other Harari architecture as typical of Harar. Exiting, I turned and looked up at the façade. On the right was a plaque with an image of a medal. This was the UNESCO prize that Harar received in 2003 for being a "city of peace and tolerance." On the left was a similar-sized plaque indicating Harar's place on the UNESCO World Heritage List. This listing came with some conditions attached, including one that the land east of Jugol be designated a "buffer zone" and remain undeveloped. Within the buffer zone, no new buildings were permitted, and the garbage dump had to be relocated. How these requirements affected the hyenas east of the town remained to be seen.

In the middle of the gate at the top, and not without irony, there was a photograph of a man in a turban: Emir Abdullahi. The emir's attire was simple, indicative of the fundamentalism to which he reverted when faced with so many threats from the outside world. And his face was characteristic of his people: a prominent forehead over almond-shaped eyes, a broad, pointed nose, and full lips. In all, it was a plain photograph of a plain-looking man, but not without symbolism. While the late emir's face in the photo remained black and white, as per the original, the background was colored blood red.

Outside Jugol, the world opened up. A broad avenue linked Andenya Menged with a highway that led from Harar all the way to Addis Ababa and the Imperial Palace. This avenue was in fact rebuilt by the Italians when they occupied Harar in the 1930s. The Italians came to Harar with a lot of optimism. Since the French-built railroad along the Rift bypassed the town in the early 1900s, Harar had fallen into obscurity; it was no longer a central point for trade between the Red Sea and the Ethiopian interior. But the Italians saw potential in the historic town and set about constructing roads and public buildings. Looking ahead of me, there was a courthouse on the right and two government buildings on the left. Each of these was built in a European style, with stucco exteriors but with a kind of colonial essence typical of buildings constructed to reflect local cultures and mitigate local temperatures.

The one closest to Harar was the Harar municipality building, responsible for services in the town; the larger building opposite was the seat of the Harari regional government. At the conclusion of the Second World War, the British handed Harar back to the Amhara, who ruled initially through the reign of Haile Selassie and then through the Dergue regime under Mengistu. Hararis suffered under the Dergue until 1991, when Mengistu was overthrown and the Harari people were granted their own regional state, the smallest in federal Ethiopia. Despite being a minority in their homeland, Hararis were guaranteed the presidency and fourteen of the thirty-six seats in the Harari state parliament. Another four seats were held by Hararis as part of a power-sharing arrangement between the Harari National League and the Oromo People's Democratic Organization.[15] Whether Harari self-determination was a good thing for the local hyenas also remained to be seen.

My hotel was a short walk downhill from the municipality building, so I headed down the road to where a disheveled-looking man guarding the gate greeted me with a broad smile. Nassir was sitting in the bar, alone among the rusted chrome tables and torn vinyl chairs. He clapped his hands when I walked in, and the clerk arrived from the kitchen with two beers. While I drank, Nassir was curious about where I'd been and to whom I'd spoken. He warned me not to deal with any other tour guides or people offering help, as they would surely lead me astray and take my money. He also asked me for fifty birr for khat, promising a *"burcha* ceremony." This is in fact an embellishing term for a group of people sitting together, chewing khat, smoking a hookah, and doing little else for an afternoon. I gave Nassir the money, whereupon he downed his beer and rushed off to the khat market in the New Town. I sat back with a beer and watched the flies weaving their paths across the laminex tabletop.

When Nassir returned with two bags of khat, he summoned me from the bar and led me across the compound to the landing under the veranda where I'd first encountered him. Some young guys were already sitting on the ground chewing khat, and I stepped over their legs to a place at the end of the veranda where Nassir had organized a mat for me to sit on. Three of the guys chewing khat were Hamar boys from the south. One, an engineering student, told me of his hopes of getting a scholarship to study in Switzerland the following year. The other two were teenagers with European sponsors. They were in Harar to enroll at a boarding school and complete tenth grade. While we talked (they all three spoke English), a Harari-Australian named Nabil emerged from a downstairs room and joined us. Nabil was born in Addis and grew up in Cairo before going to Australia at age eleven. He introduced himself as Billy and told me that this was only his second visit to Harar, so he was almost as unfamiliar with the layout of the town as I. Nassir summoned the chambermaid to prepare a clay pot with burning charcoal, upon which she put some aromatic bark. The fragrant smoke was welcome, as it drove off the flies, and we all sat chewing khat and becoming steadily more silent as the effects set in. I felt a sense of deeply focused well-being sweep over me as I watched the sun sink lower until it finally disappeared behind the hills to the west. It was time to see the hyena feeding.

Billy and two of the Hamar boys asked to come with us, so we all crammed into a Bajaj and careered down Andenya Menged, dodging crowds of people who were strolling along the road in the dim yellow of the street lights. There were few other Bajaj taxis around, and people waiting by the side of the road yelled out to our driver, entreating him to stop, or at least to slow down, and take them on. He ignored them and raced past, listening instead to Nassir, who was negotiating a price as we went. As we crested the hill in front of Jugol Hospital, the driver switched off his engine, to save fuel, and we coasted downhill through the crowds and without headlights to the gate at Argobberi. Once there, the driver restarted the engine so as to navigate a dip in the road where roadwork had caused some subsidence. The power was out at that end of town, so all that illuminated the scene was the orange glow from the burning torches in front of the market traders sitting beside the road. It was quite beautiful. We rounded a bend and headed downhill once again, at which point I could see in the distance a man sitting on the ground surrounded by hulking creatures, their long muscular necks, broad pointed ears, and dark spots evoking in me a sense of primal familiarity. Unmistakably hyenas. The headlights from a single Bajaj illuminated the scene, casting shadows far across the ground, making the hyenas look even larger. And beyond this scene—utter darkness.

Again the driver switched off his lights and we coasted down the hill. At the bottom, he veered off the road and parked under the giant fig tree, next to the other Bajaj, seemingly pleased that he wouldn't be required to start his engine and cast his headlights across the scene. He remained in the taxi, tinkering with his cell phone, while the rest of us got out and went over to the step of the shrine. The hyena man was carrying his bucket of scraps back to the shrine, all the while watched intently by the hyenas. I counted at least five adults and two subadults, one of whom was quite small. She was also the boldest and stood directly in front of the hyena man as he took his seat at the step before the shrine. The two European tourists who had just watched the feeding sat at the shrine, intending to watch the hyenas some more, but their guide urged them back into the taxi, telling them that the driver would not wait. They reluctantly boarded and headed off up the hill, leaving us in darkness, surrounded by shadowy creatures who combed the ground for scraps of meat that had gone uneaten.

Nassir introduced me to Yusuf, the hyena man. He was a tall, dark-skinned Oromo man of about fifty years with prominent, almost sharpened, features, and he looked like he had not a gram of excess fat on his body. Yusuf listened without expression as Nassir explained in Oromo what I was doing in Harar and suggested the terms under which I could spend time at the feeding place. Yusuf responded with a shrug of his shoulders, then picked up his bucket and went across to the open space, followed by the hyenas. Nassir turned to me and said, "No problem. You just pay for this time and after that come anytime for nothing." I realized only later what a generous deal this was; at the time it simply seemed fair, so I agreed. Nassir went over to our driver and after much discussion persuaded him to start his engine and switch on his headlights so we could see the hyenas feeding. He then came over to me and said I'd have to pay the driver an extra ten birr for the fuel. We sat back in front of the shrine, under the canopy of the fig tree, and watched the hyenas being fed.

Yusuf was not much of a showman. He simply sat on the ground pulling scraps of meat from a dilapidated plastic container, draping them over a stick, and offering them to the hyenas as he called their names. Tukwondilli! Baby! Dibbey! Then again, Baby! I soon worked out that Baby was the small, dark hyena who had taken up a position directly in front of Yusuf. Every now and then, Baby looked across to where we were sitting, but by and large she focused her attention on Yusuf and got the majority of the food. As the feeding continued, some of the others became bolder, and soon there was a crowd of large hyenas all vying for the bits of food that Yusuf was draping over the stick, now held between his teeth. Seeing Yusuf face to face with the hyenas like that made me realize how large their heads were. Actually, "huge" would be a more appropriate adjective. Their broad teeth flashed white from behind dark lips, their massive jaw muscles bulging beneath their cheeks. Amid all this biting and tearing apparatus, their large, placid brown eyes made for a striking contrast. While I had expected them to be at best unattractive, it came as a surprise to find them quite beautiful and engaging.

Yusuf called out to Nassir, who translated that we could participate. I declined, but Billy and the Hamar boys took turns standing beside Yusuf and holding the stick while the hyena man draped food over it for the hyenas to take. Again, Baby was boldest; she remained in front of Yusuf while the other hyenas moved away at the approach of these novel persons. Billy insisted that I photograph him feeding a hyena, but the hyenas were not being cooperative. Whenever I approached with my camera, they moved away, making it hard to get close enough to take a decent photo. Eventually, I managed a photo of Baby taking food from the stick while Billy self-consciously faced the camera. He was keen to have a photo of himself and the hyenas to show his friends back home in Australia. The two Hamar boys were excited and nervous. They had never been so close to hyenas—in the Omo Valley, hyenas either attack people or run away—so it was an exhilarating experience for both. Yusuf gave them a few more turns at feeding and then he stood up with his bucket and returned to the shrine and took a seat. He and Nassir exchanged a few words, I thanked him and paid the money, and then Nassir summoned us back to the taxi, which by that time was sitting silent and dark. We climbed aboard, the driver started his engine, and off we went, climbing up the hill to the other side of Jugol and back to the hotel.

For anyone needing to stay awake for an entire night, I highly recommend a combination of jet lag and khat. That night, I lay wide awake on my bed, staring through the darkness at the dilapidated ceiling fan, bouncing negative thoughts around in my head. What kind of meaningful observations could I ever make at that feeding place? It seemed like a plain and simple tourist operation with little to be gleaned other than perhaps tourists' opinions about the hyenas. And how were observations going to be possible when the hyenas constantly disappeared into the darkness? I had my night-vision scope, but it had limited range, and without food bribes the hyenas would never allow me to come close enough to use it. Nor had I seen any hyenas around Jugol or anywhere else in the town other than the feeding place. Was this all there was to the relationship? Just a guy feeding hyenas for the benefit of paying tourists? I had come to research a doctoral thesis, but it seemed like a big ask to squeeze an entire thesis from so little. As I pondered these questions, searching halfheartedly for a reason to be in Harar and talking myself into going back to Australia at the earliest opportunity, I heard dogs barking outside the hotel and thought I should get up to see what they were barking at. This was a challenge, though; I was in the kind of mindset where getting out of bed was almost as impossible as sleep. But then I heard a loud "oowhoop" among all the barking. Another oowhoop and I was out of bed and putting on my clothes. Another oowhoop and a lot of barking and I was rummaging through my belongings looking for my night-vision scope. I left my room and rushed along the hall to the landing outside the building, from which I could see across to the bare-earth common that the locals used as a playing field. At the other side of the common, in a sea of total darkness, a little pool of light emanated from a butcher shop. I made out seven hyenas standing in front of the shop, looking intently at someone behind the counter whom I could not see because of the side wall. The hyenas' faces were illuminated by the light from the shop, their shadows cast long across the ground. As I watched through my scope I saw bits of meat and bone flying from the other side of the counter to the waiting hyenas, who snapped up the food while frustrated dogs milled about. It was 3:30 A.M. and there was not a tourist in sight. I had my reason for being in Harar.

# 2

## LINES OF REASON FOR HYENAS

The following day, I moved out of the Tewodros and checked in to the Ras Hotel, a little way up the road from Jugol. Looking down the dim, gray-painted corridors and across bare cement floors, I agreed entirely with the description of the Ras in the Lonely Planet guide: "a cross between a boarding school and a psychiatric institution."[1] At the time, though, I thought that a little institutionalization could be just what I needed. Besides, the rooms had water, albeit a mere trickle from a cut-off pipe that only trickled between 6:00 and 9:00 A.M. Water, it turned out, was at a premium in Harar. There was a pipeline under construction that would one day draw water from an aquifer in Dire Dawa and send it via a pumping station sixty kilometers and eleven hundred meters higher up the mountains to Harar, but that seemed a distant, optimistic dream. In the meantime, tanker trucks came most Saturdays and refilled water tanks at various locations around the town. At the same time, lines of yellow plastic jerry cans snaked through the lanes, their owners queuing for their weekly share of the water. Thus I was not a little chuffed to have mine running through a small, wall-mounted water heater and emerging lukewarm from the cut-off pipe. There was also sufficient water for filling a large bucket that I could use to flush the toilet. Luxury.

There is an old saying in Harar: "The hyena who has once found food will return to that place every day for a week in the hope of finding more." After my move to the Ras, Nassir arrived every morning and waited outside the hotel for me. On the one hand, he was extremely helpful, having organized for me a phone sim and a mobile internet account, and having taken me to Dire Dawa, albeit for a fruitless attempt to obtain a residence permit. But he was beginning to act like a jealous lover, demanding to know where I'd been each day and who I was talking to whenever I received a phone call. I felt that I was sufficiently established to be able to go it alone, so I paid him for his time and told him I no longer needed a guide. Nassir was obviously used to dealing with rejection; being rejected is how a tour guide in Harar spends 90 percent of his working day. He thanked me and headed off in the direction of the khat market, leaving me sitting on the terrace of the Ras with a beer. I watched a boy walking by in the company of three baboons, one riding on his back and the other two walking alongside, holding his hands.

On my third day in Harar I had to run an errand for an American researcher living in Japan who researched Harari basketry. She'd sent me a proof of a journal article and asked if I could take it to her co-author in Harar, Abdulmuheimen Abdulnassir. While she warned that this man could be difficult and testy, she also suggested that he could be a great help with my research, as he was a store of knowledge of Harari folklore and traditions. His travel agency was almost right behind the Ras, so I took the proof with me and paid a visit.

Walking in, I was greeted by walls crowded with artifacts, postcards, flags, and posters. I called out, inquiring if anyone was in, and a voice behind a tall display cabinet called out in reply. On the other side of the cabinet, Abdulmuheimen stood and stepped from behind his desk to greet me. He was about sixty years old and physically fit, with a full beard and a mass of curly hair unsuccessfully kept in check by a baseball cap. He wore a T-shirt and off-the-shelf spectacles through which he peered with cloudy pupils that betrayed his degenerating eyesight. I introduced myself and told him why I was in Harar. "Great," he said, "this is something that really needs doing." He often stammered out his words as though they stuck in his throat. "You, you know about Ashura?" I had read about this festival, where porridge was fed to hyenas and predictions were made based on how much they ate, and my Harari contact in Australia had told me as much. "Well, you know there's another, for Aw Aslahddin. You have to find out about this." I took out my notebook and began scribbling hastily as he spoke. "Hyenas are transmitting stations for the *awliach*, the saints, you know." (I didn't know that one.) "They bring news for the people, telling them don't do this or that. That's why they don't attack people in Harar. Go to Jijiga or Awaday or some other place and the hyenas are dangerous. They eat you. But here they're peaceful because of the relationship with the awliach. Hararis, they domesticated three things: coffee, khat and hyenas. Okay, so you take that and ask me anything you need to know, but right now I have to go. We'll meet later and you can ask." I agreed to that, suggesting that I call him the following week. My American researcher friend had told me of Abdulmuheimen's failing eyesight, so I offered to read the paper to him. He nodded profusely as I read and approved the proof. We said good-bye and I went on my way with a score of questions going around in my head.

That night, I hooked up with Billy and another visiting Harari-Australian called Mohammed. We had dinner at one of the upmarket restaurants in the town: Hirut. This place had its own generator, an electric hand dryer in the toilet, and lengths of toilet paper for napkins. There was also a doorman whom Billy engaged in conversation while we waited for a Bajaj to take us home. The door-man had inquired about me, so Billy was explaining why I was in Harar, while Mohammed translated. "You should be interested in this, Marcus: two months ago, a hyena went and took a baby from in front of Selassie.[2] That's a church. The mother was this homeless woman sleeping on the street. He is saying the baby rolled out of her arms and the hyena took it and ate it down to the hands and feet." What was the response? Billy asked the doorman, who shrugged his shoulders. "Nothing. What can you do?" A Bajaj arrived and we climbed in. As we drove, Billy said emphatically, "That hyena that ate the baby was not from Harar. It was from some other place and just came to Harar. It was not from around here." Yet this was the first he'd heard of the incident. The Bajaj rounded a bend a hundred meters beyond the restaurant, and there was Selassie: a large Orthodox church surrounded by a high wall. Outside the wall and on both sides of the road, rows of prostrate figures lay sleeping under old tarps and torn blankets, giving the place the appearance of an impoverished, temporary morgue. I wondered at the fate of that baby, had a hyena not taken its life.

Waking up in the middle of the night, I left the hotel and went to the playing field where I'd seen the food being thrown to hyenas from the butcher shop. When I arrived, the butcher shop had not yet opened, but staring into the darkness I could discern the shapes of a couple of hyenas lying in the middle of the common. I sat on a stone slab and watched as dogs sniffed around looking for scraps. The dogs were taking care not to come too close to the hyenas. Not long after I arrived, I heard a chorus of barking behind me and turned to see some dogs running out of the way of a procession of six hyenas who had arrived via a side road. At the entrance to the field the leading hyena stopped, turned to watch the other hyenas coming in, and then followed the last of the group, as if checking to see that ev-eryone had followed. The hyenas began sniffing around, looking for scraps and investigating the large orange dumpster at the far end of the open area. It occurred to me that it could have been one of this group who took the baby two months earlier. The dogs were quite busy, dodging about and occasionally combining to bark madly at a hyena or chase one off. Sometimes a man crossing the field passed within a few meters of where a hyena lay without eliciting any reaction. At about

3:30 A.M., a truck arrived at the butcher shop and the door opened. The hyenas and dogs crowded around the shop while a butcher emerged from inside to take delivery of the beef, mutton, and goat on the bone. After passing the goods to his apprentice inside, the butcher emerged with some fatty strips that he held out for the hyenas, who were waiting expectantly.

The butcher's name was Sisay Gadissa, an Oromo man originally from west of the Rift. He was tubby, with a broad, gleaming smile, and was remarkably jovial considering the hour. Sisay spoke enough English to be able to tell me about the hyenas, saying he'd been feeding them in front of his shop for four years.[3] When I asked him why, he said, "I love animals. All animals." Still, he seemed to favor the hyenas over the dogs and cats. Or maybe it was simply that the dogs and cats didn't have the nerve to stand in front of the hyenas. I asked if Sisay had names for the hyenas. "Oh yes," he replied. "There's Bajaj and Gooncha. Bajaj has three legs." A broad smile briefly pushed aside his chubby cheeks. "He comes here, but today he's absent." As we talked, hyenas gathered around expectantly. Sisay went back inside and came out with a handful of scraps, which he let go of only as the hyenas took them.[4] Yes, a person can love animals, and yes, a person can want to feed them. But why not just leave a pile of food at the other side of the common for the hyenas to eat undisturbed? I didn't ask that question of Sisay, because the answer is at once too obvious and too elusive. It's not enough just knowing that they're out there. We want them to come to us and acknowledge our presence in the world as fellow creatures, but we don't know why. For some reason, we crave these creatures' validation and acceptance, grounded in their close presence. With regard to animals, the word *love* is often interchangeable with the word *need*.

The ecologist and philosopher Paul Shepard grappled with this issue. He was interested in the importance of animals to normal human development and psychological health. According to Shepard, pets, domesticates, and zoo animals are psychological band-aids for urbanites who crave the mystical presence of wild Others in their brick and lawn landscapes. The principle of phylogenetic probity holds that the healthy development of an organ is best assured under the circumstances in which that organ evolved.[5] Consequently, normal human development is retarded in urbanized humans who are far removed from the Paleolithic landscape that teemed with wild animal Others.[6] According to Shepard, *Homo industrialis* is trapped in a perpetual adolescence, "subjected to the myths of the animal/machine, heroes of progress and domination and the dualisms of ideology."[7] Hence we reach out to wild animals, wanting to draw them closer. We chase them about in boats, hike for hours in forested mountains, point cameras at them from tour vehicles for a price equal to the daily wages of an entire village, and hold out strips of food to draw them ever closer—to their peril. All in the name of filling an intangible psychological hole that emerges out of a lifeway that both facilitates and necessitates these kinds of practices. The playing field was a drab, dusty bowl amid tin and mud-brick suburbs, littered with the refuse of modernity and resounding with the chug of diesel engines. Is it any wonder that an eighty-kilogram predator amid all that should draw a man from his occupation?[8]

Later in the week, I found an apartment to rent in Jugol. A tour guide named Abdul-Aziz approached me and told me about a place for rent in the Assumberi quarter. He was a French-speaking Somali man from Djibouti with an unrestrained smile. Despite being dragged off by the police for questioning at one point, while in the possession of my one year's advance rent, he eventually came through with a nice apartment. The place was a converted upstairs storage room above a traditional Harari house. There was a bedroom, living room, and small burcha room for entertaining, all with carpet. Downstairs, in the compound, there was a shared bathroom with a real flushing toilet and hand shower. I felt extremely lucky to have found such a place in a town where even the most opulent of houses often lacked running water.[9] What's more, I could pay a monthly fee for food and the downstairs maid would prepare my meals.

On my last night at the Ras Hotel, Abdul-Aziz showed up drunk and in the company of a French-Harari woman in her thirties named Yasmin. She was attractive, with quintessential Harari almond eyes and carefully managed uncovered hair. She spoke excellent English with an accent that could have been either French or Harari, such were the similarities. They had both been out on the town and must have decided that it was a great idea to find me at the Ras and tell me all about hyenas. We sat on the terrace, and while Abdul-Aziz plunged deeper into drunkenness, Yasmin told me of her childhood in Harar. She told me of a woman who walked to work in Harar every morning and of how this woman gave food to a hyena near the stream beside the khat market at Assumberi. "Every morning the hyena was waiting for this woman, and every morning she was throwing him some piece of food. Then one morning, the woman, she stops giving food to the hyena. Instead she just walking past." Yasmin paused and looked at me, as though I needed time to understand the gravity of what the woman had done. "After this, the hyena waited there every morning, but every morning after that, the woman just walking past and not giving him any food. Like this every day for one year, and then one morning, the hyena attacked her. I was ten years old at that time and watching from my window the hyena biting here and here." She indicated her arms. "People were trying to get the hyena off, but they couldn't do it. Then somebody came with a gun and they shot him. The hyena. The woman, she was badly hurt and they took her to the hospital. The doctors wanted to treat her, but she said no. She said she did this thing to herself for not giving food to the hyena. After that she died."

I couldn't help but read a moral into Yasmin's story: establishing relations with hyenas through feeding is a prickly path to follow. Once on it, you find it very hard to extricate yourself; a hungry hyena makes a demand that is very hard to refuse. And blameless is the hyena who bites the hand that once offered food but now withholds.

When I moved into my new apartment, I was immediately invited downstairs to have dinner with the family. My landlady, Anisa, was a portly woman who was a regional manager for the national telecommunications company. She spent her evenings lying on one of the platforms in the main house, calling out to the maid, Medina, to bring various snacks and drinks from the hut in the compound that served as a kitchen. Anisa complained to me that Medina was too simple and too slow but that she had to keep her on because Grandma liked her. Grandma was Uma Safia. She had a broad smile, husky voice, and pale blue eyes that seemed incongruous in Ethiopia. She spent her days in the *kirtat*—the room to the side of the living room—sitting cross-legged on a platform and chatting with Medina, who sat on the floor doing her various tasks: shelling peanuts, sifting lentils, making coffee.

Although Anisa was among the wealthier women in Harar, in one way she was considered impoverished: she had no children. The only other member of the household was Anisa's partner, Faysal. He was short and stocky, and his skin was dark from time spent in the sun working on his hobby farm. Faysal also liked khat a lot and chewed from midafternoon until late into the night. Still, he was awake by 7:00 A.M., when he summoned me from my apartment so he could take me to visit his farm. First we had to go to Assumberi to buy khat and organize a Bajaj to take us out of Jugol and into the countryside. At the khat market, we met up with two guys who would accompany us. One was Tofiq the farmhand, a tall, muscular Oromo man who looked like he could carry all three of us to the farm himself. The other was Adil, a short Harari man with Arabic features who spoke excellent English. We rode the Bajaj down a bumpy dirt road past Argobberi, out into the farmland and to a small stream two kilometers beyond the cemetery. We got out and followed a path beside the stream through groves of banana trees and rows of khat to Faysal's farm. It was a lush couple of acres brimming with khat, peanuts, and sweet potatoes thriving amid mango, coffee, and guava trees. Tofiq set about digging a furrow while Faysal, Adil, and I sat beneath a tarp strung between two trees, where they told me about hyenas.

Faysal was not wholly confident with his English, so he spoke in Harari while Adil translated. "The hyenas can be a big problem for farmers; they step on the maize when they are playing or having sex. You can't do anything. If you harm the hyenas, they take revenge; they destroy your farm and your animals. But if you don't hurt them, they leave you alone. They even keep other hyenas away from your farm; they protect you." Faysal went silent for a while as he chewed his khat, and Adil took up the lesson. "When I was a boy, out walking with my family, if we came across a hyena, we say *derma sheikh*, and he let us pass, no problem." I knew about this name, derma sheikh. It translates as "young man of religion." But I needed to find out more. Why did you say derma sheikh? "It encourages him to be peaceful, to let you go in peace." As though that was enough information, Adil moved on. "There was this man at Suqutatberi who knew hyena language. He didn't go out after 6:00 P.M. because the hyenas knew him. They killed him." My handwriting degenerated quickly as I scribbled furiously in my notebook. Adil took no heed of my writing speed, or lack thereof, and immediately went on, "There is two groups of hyenas in Harar, one at Assumberi and one at Argobberi. The female hyena from Assumberi, they come to Argobberi to take the males. These times there is a lot of fighting between the hyenas. This is at Aw Abdul, that mosque we passed outside Argobberi." A picture was emerging of a very subjective hyena, socially engaged with the Hararis; and this was born not of ignorance but of close association with and attention to hyenas.

Tofiq was digging methodically nearby, exhaling each breath with a rhythmic hiss—an Oromo mannerism. Adil went on, "There is this kids' saying, *waraba wolada, Gungora salata*. It means "the hyena gives birth, the leopard is saved." They sing it when there is a certain kind of rain. It is because hyenas' babies live in holes beside the river, so the flood kills them, but the leopard lives in a tree and survives." He thought for a moment. "This is why it's dangerous to see a hyena when it's raining. The house will be flooded and the hyena will be distressed. Very dangerous at these times." I asked whether hyenas were dangerous at other times. "Not in Harar, but in another place, very dangerous. When I was working in Borayn, in Addis, I had to go home in the dark because of night shift. I was so scared of hyenas because in Addis the hyenas will kill you. The other workers, they went home this way and I, that way by myself. I asked to be taken off night shift but they said no, so I left my job. This is because of hyenas. There was this drunk man in Borayn who slept in the street. He always said he would be eaten by hyenas. It happened." Faysal, who was focused on his khat, changed the subject. "When people find a dead hyena, they cut off the . . . this part"; he indicated his eyebrow. "They make a *kertass*, a bracelet for the arm, up here, and it's protection from, you know, the evil eye." Adil finished, saying, "That's the Amhara people. For Hararis, it's different. For us, the hyena skin makes you brave." It was an enlightening few hours we spent beneath the tarp, and I had a lot going around in my head as we walked the road back to Jugol in the afternoon. My spinning thoughts were combined with dizziness, which I thought at the time was due to the altitude but turned out to be giardia. I discovered only later that whenever Medina cooked pasta or rice, she gave it a good rinse in town water before serving it up.

The day after my visit to the farm, I had an appointment with Abdulmuheimen. He'd canceled our first appointment, saying he was overtired, so I was relieved when I arrived at his travel agency with a bag of khat and found him behind his desk, raring to go with the interview. I handed him the khat and he reprimanded me, saying that I should not bring him anything in the future. Then he told me to have a seat and asked if I had found a place to live. I told him about my apartment in Jugol. "That's good. If you didn't live in Jugol, I wouldn't talk to you." I wondered whether he meant that. After listening to my ethics statement and giving me permission to record our interview, he lit a cigarette, and away he went. "Donkey was walking along the road"—Abdulmuheimen bobbed his head to mimic a walking donkey—"when he met Hyena. Hyena looked up at Donkey and said, 'Hey, are those horns on your head?' Oh, Donkey was afraid, and he thought, 'I can fool Hyena and scare him with the horns,' so he said, 'yeah,' and nodded like this. But the ears [were] like this waving around, so Hyena ate him up." My first Harari folktale. "Mohammed Gragn sent a message to the Christian emperor: 'I am as strong as a lion, as fast as a hyena, as cruel as a mother. I am coming.' When the emperor asked how a mother could be cruel, the messenger said she would do anything for her baby."

You said that hyenas are transmitting stations for the saints. Can you tell me what this means? "Sure. The word for hyena, *waraba*, means newsman. *Wahr* means news and *abba*, it means man. Waraba is the newsman for Hararis; they bring the news from the awliach, the saints.[10] There is people in Harar who know what the hyenas are saying. The people take the news and give it to the Hararis. You want to understand hyenas like them? You kill a goat and hang up the skin, like this, with the rocks in it. The hyena comes and eats the skin; then you chew it, and you can understand the hyena after that. But don't do this, because the secret will pass and the hyenas will attack you." I remembered Adil's story of the man who knew hyena language and couldn't go out after 6:00 P.M.

Can you tell me why people say "derma sheikh" to hyenas? "First let us define the term derma sheikh. This is a combination of two words, usually used as one word. Now, you, Marcus, should be called waraba sheikh because you are studying about waraba; then you are teaching people. Derma sheikh means, *derma* is youth, *sheikh* is teacher. Derma sheikh means those who are devoted for religious life. Derma sheikh means a young person who is not yet married but he is devoted to his religion. As for the hyena, why do they call him derma sheikh? It means that this hyena, due to his environment, he became devoted for his faith. His ethic is not to harm people. The hyena of Harar. We are symbolizing them with those who are devoted for the youth because youth is troublesome. He's making problems. So that derma sheikh also symbolizes [that] the hyena is a wild animal who will attack you in another place, but not in Harar. That's why derma sheikh. He's peaceful.

"Abadir was sitting, facing the Kaaba [in Mecca], and a hyena came running along from the east to this way, the west. He took a toothbrush from his pocket and threw it at the hyena, hitting him, breaking the leg. So now Harari people see a hyena, they say Abadir *hokola*. Means, Abadir domesticated."

An origin myth featuring Abadir and a hyena! I was compelled to interrupt before Abdulmuheimen shifted topic. Why call the hyena Abadir? Is he a manifestation of Abadir? "No. Abadir is anybody. Child of Harar. You grow up in Assumberi, Argobberi, someplace, you say, I am Abadir. Anybody in Harar. You Marcus, you are *gey woldi*, child of Harar. You stay, learn Harari, chew khat, living in Assumberi. Then you can say, 'I am Abadir.' Hokola means like this"—Abdulmuheimen mimicked a limp. Lame? "Yes. He can't run, so he's hokola, domesticated. Three things Hararis domesticated: coffee, khat, and hyenas. Abadir made the hyena peaceful, and anybody comes into Harar will be domesticated." So what's the significance of the toothbrush? (In Harar, a toothbrush is a stick from a particular type of tree. The bark is peeled from one end, which is then inserted into the mouth and swiveled about to clean the teeth. Men often twirl their toothbrushes in their mouths as they walk.) "No significance. It's just convenient. You keep it here [chest pocket] and you need to throw something, you have nothing else. Hararis are peaceful, so they don't take a knife. They have a toothbrush." To me, that seemed pretty significant. "Hyena woke up in the morning and was running to Harar to attack the people. But he found the garbage and the ash, and he ate that. *Waraba min be yiburdal? Hammad be*. With what is hyena satisfied? With ash."[11]

I was visiting the hyena feeding every night and sitting at the shrine while tourists and hyenas came and went. Yusuf was quite aloof, and during the nonfeeding times he sat silently at the shrine. He spoke no English and I spoke no Oromo and little Harari, so communication was minimal. His son, Abbas, was more engaging. Sixteen years old and as lithe as his father but with a chubbier face, Abbas was cheeky and recalcitrant. He often fed the hyenas while his father was in the house, and he was nowhere near as tolerant of difficult tourists and tour guides as his father was. If people wanted some kind of deal or special price, Abbas simply rebuffed them, and he laughed if they threatened to take their business to the other feeding place. Indeed, it was almost mischievous the way that Abbas stonewalled people and watched their resulting indignation with a broad smile. But he also enjoyed sharing the joke with me, and after having upset someone, he returned to the seat of the shrine, sat beside me, and elbowed me, while pointing and remarking on another success. I liked Abbas a lot.

One of the guides whom Abbas took great pleasure in foiling was Tusty, a handsome young Oromo guy who spoke quite good English but didn't have a tour guide license. Tusty's typical approach to drumming up business was to wait around at the feeding place for unaccompanied farenjis to arrive and then try to persuade them to hire him. On some occasions, he ran down the road ahead of a vehicle full of farenjis and directed them to the feeding place as though he was showing them the way. Once the tourists were out of the car, Tusty stood beside them, explaining the hyena feeding and mediating between the farenjis and Yusuf, giving the impression that he was their guide. Then, after the tourists departed, Tusty went to Yusuf for his percentage of the fee that the tourists had paid. Abbas was not as trusting of Tusty as Yusuf was and operated on the assumption that Tusty was pulling a scam. So when Tusty brought bona fide customers and Abbas refused his commission, loud and protracted arguments ensued. Tusty quickly cottoned on to what I was doing in Harar and offered to take me out into the farmland in the morning to show me where Yusuf's hyenas had their dens. I didn't have to pay unless his information was good, so I agreed.

We set off at 7:00 A.M. for the farmland to the east of Harar. Tusty led me along a dirt road past groves of eucalyptus, across moorland, and downhill to where the road dipped into a deep cutting. I wondered how many feet, hooves, and paws had trampled the soil to cut the road so deep. There was a stream at the bottom of the cutting, but instead of crossing, Tusty led me along a narrow, overgrown path that ran along the top of the bank. After walking fifty meters, our legs were soaked by the dew on the bushes that crowded around us. We reached a small crest whence Tusty indicated an impenetrable sea of rushes and sugarcane through which it was impossible to see anything. "That is the hyenas' house, that place there." Possibly he read the disappointment in my expression, so he went on with some enthusiasm. "The other hyena place—the Assumberi hyenas—that place is easy. You can see there and it's very close to Jugol." Sounding a lot like a child who'd been taken to a B-grade fairground, I protested, You said I could see hyenas at this place! "Yes, but it's growing here. There is too much of the forest. You have to go down but you can't." I agreed that he would show me the other hyena dens the following day and told him he could go back to Jugol while I hung around beside the stream for a while. I found a place above the cutting where the stream crossed the road and sat on a rock outcrop, staring across at the farmland, looking for signs of hyenas. After about twenty minutes, I saw some movement between the mango trees. Then a flash of spots and a hyena emerged, heading at a lope toward the very place that Tusty had indicated as the hyena dens. Well done, Tusty. I wasn't going to see much happening at the dens, but at least I knew where they were and which way Yusuf's hyenas would be heading home each morning.

The following morning I met up with Tusty again, and he took me north of town via Gunda Fera, the suburb outside Harar that was once a leper colony. Again we descended toward a stream, which we followed to the bottom of a rock outcrop rising up forty meters. Tusty led me along a path between rocks and pointed out all kinds of evidence of hyenas—trampled cactus, burrows between rock crevices, and chalky white hyena feces. We crossed in front of the outcrop to a place just above and behind it where there was more open space from which I could make observations. The rocky outcrop was ideal for hyenas. It gave them plenty of cover and ample places to dig out burrows, and it was only two hundred meters from Jugol, and just a little further from the feeding place to the north. It also provided me with a secluded spot to sit and make observations with a clear view of the dens and not a lot of disturbance from people. Under the circumstances, it was as close to perfect as it could be. Tusty explained, "These hyenas, these Assumberi hyenas, they don't feed from Yusuf; they feed at the other hyena man. He is Mulugayta." I was beginning to think that I should focus my efforts on this northern clan. It would certainly be easier to make observations at the den. I asked Tusty if he could introduce me to the other hyena man. He agreed and led me down a bare granite slope and up a steep path to a road that took us past the slaughter yards. Walking between the wall of the slaughter yards, a stone building, and some huts, Tusty held out his arms and said, "This is hyena-feeding place." It wasn't a very broad area, just enough for a truck to turn around in. We continued on past the stone building to a gate and entered a compound.

The house within the compound was freshly painted and well kept. Tusty knocked on the door and continued until the door opened.[12] We were greeted by a heavyset, shaven-headed Amhara man wearing a blue singlet, track pants, and a considerable amount of jewelry. Grinning broadly, he extended a hand to me, and Tusty introduced him as Mulugayta the hyena man. Tusty explained to him in Amharic what I was doing in Harar and that I wanted to spend time at the hyena-feeding place making observations. They discussed price and Tusty turned to me, saying, "No problem. You can pay three hundred birr and come as long as you like." I was uncomfortable paying Mulugayta more than I paid Yusuf, so I told them I'd think about it. Walking back to Jugol, Tusty pointed to a purple Peugeot at the side of the road and said, "This is his car." It was the only car there.

# 3

# BETWEEN DIFFERENT RELATIONS

I wanted to find out more about this practice of hyena feeding in Harar, so I arranged to meet up with a local journalist named Amir Ali Aqil. His office was a surreal sort of affair, a construction on the roof of the Tourism, Sport, and Culture Bureau, possibly the highest point in Jugol. Two sides were walled, while the other two were open to the elements, making paperweights an absolute necessity. We sat down on either side of his desk, and while kites and vultures swirled about behind our backs, Amir told me what he knew about hyena feeding. "Did you know Yusuf was attacked by a hyena when he was a child? But his father didn't feed hyenas; Yusuf was the first of his family, not third-generation like people say." When did hyena feeding begin in Harar? "Tour guides tell farenjis it's an old tradition in Harar, but that's not true. There has only been hyena feeding in Harar since sixty years. The first man was Aw Dudzo from Suqutatberi.[1] He started by feeding skin to hyenas outside the wall. After that there have been many others." As an afterthought, he added, "a total of fifteen people." That seemed quite a high turnover for a period of sixty years, and I wondered about both the accuracy of the information and the reasons for such a high number. When I was about to finish up and leave, Amir volunteered, "Hey, do you know about Ashura?" I nodded. "You can still see this at the shrines Aw Aboker, Aw Nugus, Aw Hakim. If the hyena eats all the porridge, then the coming year will be famine, war, and drought. If he eats and is satisfied and some left, then a good year." How much did he eat last year? "I don't know. I've never seen it."

I went to the Arthur Rimbaud Museum in the center of Harar to look over the collection of articles in the downstairs library.[2] Among one of the volumes, I found an article written by Sylvia Pankhurst for the *Ethiopia Observer*, dated March 1958. The title: "Dinner Party with the Hyenas." In her article, Pankhurst describes visiting Harar and being driven east across a moor and through a cutting to a place where an Oromo man named Mohammed Hassan fed hyenas inside his hut. The hyenas were skittish at first, but as the night progressed they became quite bold and the experience intensified. This is the earliest written account of hyena feeding near Harar.[3] In Wolf Leslau's compilation of Harari cultural practices in the 1950s, hyena feeding isn't mentioned. But then, Leslau's Harari contributors were attending Addis Ababa University at the time, so they may not have known that someone had begun feeding hyenas in Harar by then. When Hans Kruuk visited in the mid-1960s, hyena feeding was being practiced right outside the wall by a man named Tana.[4] By chance, I met a Harari sports journalist, a short, silver-haired man named Mohammed Sayeed who'd lived his entire life at Suqutatberi. He remembered Tana and a few others. Mohammed told me that Yusuf started feeding around 1994, which is the time of Camilla Gibb's fieldwork, and said that Yusuf used to be stingy with the food. But something happened and Yusuf changed his ways, giving food to hyenas regardless of whether tourists were present. As for Mulugayta, he started feeding in 2005–6 specifically to earn an income from tourism.

I needed names for the two clans of hyenas that were feeding in Harar. Some of the locals called them the "Muslim" and "Christian" hyenas based on the religions of the respective hyena men, but it would have been odd ascribing a religion to a group of hyenas, even if only by name. I recalled Abdulmuheimen saying that the hyena territories around Harar were defined by the shrines, so I decided to name them after two shrines that lay in the directions of the two clans' territories: Aw Sofi and Aw Aboker. The Sofi clan would be the group of hyenas who lived in the dense vegetation beside the stream to the east of Jugol and who fed from Yusuf; the Aboker clan, the hyenas who occupied the rocky outcrop to the north of Jugol and fed from Mulugayta. I guessed that it was Aboker hyenas who were congregating at the playing field and being fed by Sisay the butcher in the early hours of the morning. This place was quite close to the Aboker feeding place and only a couple of hundred meters upstream from the Aboker dens. It was difficult to imagine yet another hyena clan in such close proximity, considering the typical territory size of hyenas in national parks. Everything was pointing toward my study having to focus on the Aboker hyenas—it was obviously going to be easier to find opportunities to make observations—so I decided to take Mulugayta's deal and spend alternate nights at the Aboker feeding place. I could make up for the differential in what I paid Yusuf with some bags of khat.

Walking to the Aboker feeding place through the crowds of people milling about at 6:30 P.M., I encountered a man pushing a wheelbarrow containing a cow's head, with about a thousand flies riding shotgun. If that was the sort of thing that the slaughter yards had on offer, then I was definitely heading toward hyena-friendly territory. I rounded a bend and found myself in a space between the slaughter yards and a stone building, where I was immediately singled out and intercepted by a small young man of about twenty years. "Stop, this is my land," he said. "You have to pay to come in." It was obviously a public road and he'd singled me out from among a score of Oromo people who were going to their farms. I figured he must have been collecting money for the hyena feeding. I told him I had a deal with his boss, Mulugayta, and that I'd come to pay the money. He paused and looked at me before hurrying off in the direction of Mulugayta's house, leaving me standing on the step of the stone building. While I waited, I looked over to the open area through which the road passed. Four hyenas lay in semidarkness beside the road near the wall of the slaughter yards. They didn't seem bothered by the people walking past within meters, a real contrast to the Sofi hyenas, who preferred to keep their distance from passersby. Mulugayta arrived and we shook hands. I gave him the money, which he counted carefully, and then I asked him to sign a receipt, which I needed for my university.[5] He called to a couple of young men who were standing beside the road and gave them some instructions before returning to his house.

The taller of the men was Binyam. He had a mat of curly hair and an unrestrained smile, and he spoke enough English to be able to introduce himself and promise to teach me Amharic. It turned out that Binyam's job was to look after the baskets of food for the hyenas. When a basket had been emptied, he had to replace it with a full one and then take away the empty to be refilled. The other of the two young men was Saloman, a thin fellow with large, gentle eyes. I recognized him from an internet video I had seen of hyena feeding in Harar. His job was to feed the hyenas while Mulugayta was indoors. Saloman invited me over to where there was a pile of cobblestones and indicated that I could sit and watch from there. The young man who had intercepted me at the entrance to the place was Balay. His job was to collect the money from the tourists and make sure nobody watched the feeding without having paid. Ignoring me, Balay went over to the step of the stone building and sat, waiting for tourists to arrive. I asked Saloman how much he was paid to feed the hyenas. He told me 10 birr per night, equivalent to about US60 cents. Meanwhile, it cost farenjis 50 birr and Ethiopian nationals 25 birr to see the feeding. Doing the math, I figured that Mulugayta was on a pretty good wicket.

Watching Saloman feeding the hyenas that night, I was struck by the outward similarity of the two feeding operations. In the same way that Yusuf did, Saloman sat on the ground or on a rock with a container of meat scraps. He draped strips of the scraps over a stick clamped between his teeth while the hyenas approached and snatched the food. Vehicle headlights illuminated the scene, and cameras flashed as tourists took photographs. As with the Sofi feeding place, tourists were invited to stand beside the hyena man and hold the stick, over which pieces of meat were slung for hyenas to take. Even the prices charged were identical. Indeed, aside from some details such as the receptacles used to hold the food, the kind of food on offer, and the number of men involved in the feeding, there was little to outwardly distinguish the two places. Then Mulugayta arrived.

Brimming with confidence and jewelry, and wearing his characteristic grin, Mulugayta strode through the line of spectators straight toward Saloman. The young hyena feeder noticed the arrival of his boss and quickly stood, handing over the food basket like a guitarist on amateur night handing his instrument to Bob Dylan. Mulugayta took the basket without acknowledging Saloman and began the show. His opening was to scatter some pieces of meat at the feet of the tourists, who stood at the periphery of the feeding area. The Aboker hyenas proved themselves far bolder than Yusuf's Sofi hyenas as they rushed toward the tourists to collect their food. Needless to say, this was at once intimidating and exciting for the tourists, who variously backed away or held their ground so as to enjoy a close encounter. Then Mulugayta began feeding. He took pieces of meat and held them between his teeth—no stick—and stood leaning over the hyenas, who reached up to snatch away the food. The vehicle engines seemed to idle more loudly, the chatter among the spectators increasing in frequency and volume. Mulugayta singled out a hyena, called her name, and held the basket in such a position as to invite her to insert her head. The hyena obliged, and while her head was buried deep inside, Mulugayta threw a leg over, straddling her as if hyena riding.

After dismounting the hyena, he sat down on the ground and invited people to participate. He singled out one attractive Amhara woman in the company of her farenji partner and told her to come and try. She adamantly insisted that she didn't want to, but after some coaxing from her partner she eventually sidled over to the hyena man with her hands held up to her chin and sat carefully on the ground. Mulugayta held out a few bits of meat to allow the woman to become accustomed to the hyenas, who were outwardly unconcerned about her presence. She sat completely immobile, hands held up to her chin, while the hyenas circled around snatching pieces of food. Then the hyena man put his arm around the woman's shoulder, holding a piece of meat very near her face. A hyena approached to take the food and the woman instinctively turned away from the hyena toward the face of Mulugayta, whose grin grew wider.[6] He wasn't even looking at the hyena. When the show was over, one of the tour guides initiated a round of applause.

Some of the tourists clapped enthusiastically, some halfheartedly, and some simply turned, boarded their tour buses, and prepared their cameras for the next attractions on their itinerary. Okay, so that was different.

During the time I spent alternating between the Aboker and Sofi feeding places, I began to notice all sorts of minor details that differentiated the two operations. The Aboker place was more carefully organized. The workers arrived at exactly 6:00 P.M. and began preparations for the night's feeding. They remained at the feeding place whether or not tourists were present and left at exactly 9:00 P.M. At that time, any food that was left over was dumped on the ground for the hyenas to set upon en masse, and the workers headed for home. The hyenas were also restricted by these times. When they weren't feeding, they typically congregated in a space between the road and one of the walls, where they lay down together and waited like performers waiting in a dressing room. They were free to wander off to the farmland at any time, and many did, but the employees prevented the hyenas from going toward Jugol. Whenever a hyena got up and started in the direction of the Old Town, the employees headed her off and threw rocks to drive her back to where the other hyenas waited patiently. Only at 9:00 P.M. could workers, hyenas, and anthropologists go their own ways. Spending time at the Aboker place felt like working in a sometimes bitterly cold factory, and I found myself watching the time in anticipation of knocking off.

The Aboker employees not only tried to restrict the hyenas' movements but used a call to attract them at the start of the feeding and sometimes during the feeding when few hyenas were present. I remember evenings walking to the Aboker place and hearing Saloman's voice ringing out across the town: *Num cha, lo, shebe shebey. Nach ho!* I don't know if the call worked, as only a few hyenas ascended the hill while Saloman was calling, and these might have arrived to feed in any case. But Saloman seemed to think that it worked, so he stood on a rock outcrop at the far end of the feeding place and called out across the valley every evening. At times when the hyenas were preoccupied with a dead livestock animal in the farmland and only one or two were at the feeding place, Mulugayta took over the feeding while Saloman went and called out to attract the hyenas, entreating them to come and entertain the customers. One evening as the sun was setting, I stood beside Saloman on the rock, watching the hyenas arriving as he called. When I shone my flashlight on a juvenile who was ascending the hill, Balay, who was standing beside me, pushed my flashlight down, afraid that the hyenas might be frightened off by the beam of light. I came to realize that these guys were acting according to a concern that the hyenas would not come. The set start and finish times, the calling, the rock throwing were all efforts to add some predictability to an operation that depended on autonomous, unpredictable hyenas. The feeders were in fact in quite a precarious position.

As for Yusuf, there was no set time for him to begin hyena-related work. He arrived home from the farm at around sunset, when the hyenas were beginning to arrive. He washed up, said his prayers, did some preparation, and then took his bucket of scraps to the shrine beside his house, where he sat and waited for hyenas and tourists to come. Nor did he have a set finish time. When it looked like no more tourists would be coming, he emptied the leftovers onto the ground for the hyenas and went inside his house, or returned to his farm to guard against khat thieves. Sometimes tourists arrived after the leftovers had been given out and went away disappointed because the hyenas were no longer feeding. Yusuf also purchased bags of bones for the hyenas. I called them "calcium supplements" because they were rarely used for entertaining tourists. Normally, Yusuf emptied the entire bag of bones onto the ground, even while there were tourists present, at which point the hyenas snatched up the choicest bones and ran off into the darkness to demolish them, hotly pursued by other hyenas who wanted what they had.[7] This was often annoying for the tourists, because the hyenas were subsequently unwilling to approach and take food from the stick. The Sofi hyenas came and went as they pleased; they often disappeared into the woods beside the feeding place to engage in altercations with dogs, or went down the road to the garbage dump and sifted through the day's bounty.

Yusuf never called out to attract hyenas at the beginning of the night; the hyenas simply arrived or didn't, depending on their whims. They knew that there was food available at the place each night, so the choice was theirs. Yusuf gave names to all the hyenas who came and fed, and he called their names when they were already present to encourage them to feed, but he rarely called out into the darkness. Like the Aboker hyenas, the Sofi hyenas occasionally found carcasses in the farmland or on the nearby road, and on these occasions they had little inclination to stand in front of a bunch of tourists. They also had the option of seeking food at the nearby garbage dump, which was deserted by people during the night. When the hyenas were absent from the feeding place and there were expectant tourists present, Yusuf went to the edge of the feeding area and whistled out into the darkness to try to attract them. If they didn't come, he apologized and refunded any money paid.

Even the smallest details spoke volumes about the differences between the two places. At the Aboker place, they used three large new baskets to hold the hyena food. Yusuf used an old plastic jerry can with the top cut off, holes pierced at the top, and a piece of string tied through the holes for a handle. The Aboker feeders gave the hyenas good-quality fleshy scraps, while the Sofi hyenas were fed mostly strips of skin with little flesh and lots of hair. When hyenas at the Aboker place gathered in the local dumpster to sift through the refuse, the employees evicted them by throwing rocks, which banged against the metal sides. When Sofi hyenas gathered in their dumpster and Yusuf had expectant tourists waiting, he just called the names louder and complained to the hyenas that they weren't doing their jobs. In my field notes, I called this the "Twister distinction," after the cheesy 1996 movie with Helen Hunt and Bill Paxton. In the movie, two storm-chaser outfits vie against each other to set their data-gathering equipment in the path of a tornado. The outfit operated by the heroes is casual, quirky, poorly funded, and ultimately successful. The outfit of the villains, by contrast, is ruth-less, impersonal, and efficient, with a fleet of black SUVs and plenty of funding. The dictates of Hollywood endings ensure that this outfit is doomed to fail and that the villains die, the inevitable consequence of having too much money and not enough heart, not enough spontaneity. But cheesy movies always give us endings we desire rather than endings that make sense in the world of things. Where life reflected art in Harar, the realities were reflected as opposites. Yusuf's operation was struggling.

Within the Harar tourism econo-system, the keystone species is definitely the tour guide. Locals refer to these people as *farenjatali*, meaning "one who lies to, para-sitizes, or makes a living off farenjis." Farenjatali is a derogatory term, but it is used with a hint of jealousy; tour guides can make a lot of money just for walking around with tourists, showing them the local attractions, mediating transactions, chasing away stone-throwing children, and fending off beggars.[8] Compared to some of the low-paid, difficult jobs in Harar, being a tour guide is pretty desirable, and all that one requires is a basic knowledge of English, a working knowledge of Harar and its attractions, and a tour guide's license. Nevertheless, a lack of these qualifications is no bar to an ambitious young man who wants to cash in on the tourism industry; I'd heard many young, unofficial guides invent facts about particular landmarks or practices in Harar.[9] The tourists didn't know any better and readily believed whatever the tour guides told them. They also followed the guides' leads with regard to which places they visited and in what order, and it was this position of control over tourists' beliefs and choices that in turn gave the tour guides control over the fortunes of the hyena-feeding places.

While I was in Harar, many of the tour guides shunned the Sofi feeding place because of the unreliability of the hyenas.[10] Their frequent absences were embar-rassing and frustrating for tour guides, as their customers arrived expecting to see hyenas feeding. The guides also resented having to wait around for the hyenas, because the sooner the tourists were finished at the feeding, the sooner they could collect their commissions and go home or out drinking. If the Sofi hyenas failed to materialize, the guides had to take their customers to the Aboker place, where the chances of seeing hyenas feeding were much better and the hyenas bolder. Sofi hyenas often hung back in the shadows while Yusuf pleaded with them to come and feed; the standoff was broken only when Yusuf conceded and threw the food over to the hyenas. This was extremely frustrating for the tourists, who wanted up-close encounters with hyenas rather than a guy throwing scraps into the darkness beyond the range of their camera flashes. The guides often complained to me about this state of affairs. "They are always late, or sometimes not there at all," complained one guide, who then extolled the virtues of the Aboker feeding place, saying, "those hyenas are always on time and are much bigger and very tame—like dogs." And so they favored the Aboker feeding place.

Is that what tourists want? Hyenas like dogs? Spending so much time at the hyena feedings, I had a lot of opportunities to ask tourists about their impressions. Two young Danish guys and a German girl said they'd visited both places and definitely preferred Yusuf's. They said it was less formal, whereas Mulugayta tried to control the hyenas too much. This was a common theme; people preferred the more relaxed setting at the Sofi feeding place, and not just because there was somewhere to sit and it wasn't as bitterly cold. I met a French couple in their mid-fifties who were traveling around Ethiopia for a couple of months. They complained about their guide; they'd asked him specifically to take them to the "old" hyena-feeding place and instead he took them to Mulugayta. They hated it and complained that it was too touristy. But perhaps I was seeking opinions that matched my own. Maybe the tourists at the Aboker place who oohed and aahed and clapped enthusiastically before getting back on their buses preferred the predictability, control, and spectacle of that place. After all, animals performing tricks in front of bleachers can draw large crowds. And as for Ethiopian tourists, these people might have seen the hyena feeding from yet another perspective. In Ethiopia, hyenas are considered deadly, dangerous animals. Nor are they consigned only to national parks and reserves; there are hyenas in Addis Ababa and Dire Dawa, and Ethiopians are genuinely afraid of them. Possibly a tame, controlled hyena is fascinating to an Ethiopian tourist as a bizarre, frightening anomaly. Tour guides often remarked to their customers, "Look, it's just like a dog." This is more a comment on the anomaly of hyenas in Harar than a way of enhancing the excitement of a close encounter with a dangerous predator. Still, I'd spoken to enough tourists to recognize that a significant number of people favored Yusuf's relaxed, noncontrolling approach to Mulugayta's razzle-dazzle. And hunting around, I found a newspaper article from the *Malaysian Insider* in which the writer expressed this very preference as well as her disappointment at seeing the Aboker feeding. The writer, Lim Ka Ea, describes her first impressions of Saloman feeding the hyenas:

> I was [too] completely engrossed with my own thoughts to even notice how close I was to the hyenas and two had in fact brushed against my legs.
>
> It was a chilling experience, sometimes I would wonder whether the hyenas might go on a rampage, but most of the time, just touched by the thought that human beings and wild animals can co-exist in harmony together, just the way nature had intended. I had tears brimming in my eyes.

Ka Ea seems to have been drawn into a mystical experience with the hyenas, but Mulugayta's appearance quickly snaps her out of her reverie.

> The tears didn't last long because unfortunately, this practice is now very much tourist-orientated.
>
> Once every last strip of meat had been devoured by the hyenas, an overweight bald man stepped in to replace Solomon [sic]. For added value, the man gave several performances to demonstrate his chivalry and dominance over the hyenas, much to my disgust although to the delight of many others. He was forcing one of the hyenas to lie on the floor so that he could sit on it. In my mind, I thought there was no chivalry at all since the hyenas were all fed by then.
>
> Adding salt to injury, in the middle of his performance, his mobile phone in his pocket went off and he answered it with a great sense of importance. For me it just ruined the whole thing because then, it just proves how men will always want to dominate things. It also made a mockery out of the whole ritual by disturbing the elements of what had supposed to be a traditional practice in tune with nature. The co-existence and co-dependency between human beings and animals are being tossed out in the name of power and greed.[11]

So what is it about Mulugayta's performance that some of us find so offensive? On the one hand, I suppose it's the contrivance of it all. When people want to feel like they're exploring the world and discovering new and interesting things, it comes as a massive letdown to find a man in a tracksuit and a bunch of well-trained hyenas doing the same old tricks. There are no surprises, no spontaneity; the hyenas won't be going on a rampage. Instead, the hyenas perform, the spectators applaud, money is paid, and taxis are ridden off into the night in search of dinner. Next!

But there's something else, and it runs deeper than the deflating, de-individualizing, realization that you're just another tourist on a well-worn trail. Note Ka Ea's use of the terms "harmony," "in tune," "co-dependency," and "co-existence." These terms speak of a desire for a more balanced encounter, an encounter in the middle ground where nobody is really controlling the situation and the participants, including the hyenas, are all there on their own terms. Whether this is achievable is beside the point; it is the desire for engaging the Other in the middle ground that is important, and this desire was being disregarded. Instead, the hyenas were being coaxed into the human realm with pieces of meat and made to conform to human needs. And so begins the slide down a slippery slope toward a world of spotlights and bleachers, handlers reeling off scripted commentary into headsets, and family tickets with set start and finish times for the performances. This is not a middle ground, not a coming together on equal terms, but the conforming of hyenas to the immediate needs of humans. As such, we are not given the opportunity to venture into their mysterious world because they themselves have been removed from it. Instead, we find ourselves in the same old human world, but a human world with shadows of hyenas.

This leads me to another question. If Ka Ea and others so desired a more balanced encounter, one in which the hyenas had a greater say about the terms of engagement, then why didn't they all go to the Sofi feeding place? As I said, the tour guides largely made the choices for them, but that is not the underlying reason. Ultimately, it was due to the actions of the hyenas and their compulsion to avoid humans. This is the paradox articulated by John Knight of Queen's University in a modest but meaningful article titled "Making Wildlife Viewable: Habituation and Attraction."[12] Knight's paradox holds that the qualities of wild animals that make them attractive to tourists—their wildness, unpredictability, and tendency to avoid humans—are the very things that must be reduced in order to make viewing possible. Tourists were coming to Harar with limited time, limited budgets, and the expectation of encounters with hyenas. These factors trumped their desire for balanced encounters with hyenas because, with only a couple of days in Harar, any encounter is preferable to none. Hence the pressure on tour guides to fulfill the tourists' needs and thus their favoring of the Aboker place, where the operators made efforts to reduce the hyenas' unreliability and their tendency to avoid humans. There they adhered to the set start and finish times, restricted the hyenas' movements, and withheld food until hyenas came to feed directly. Meanwhile, Yusuf was nowhere near as intent on controlling the hyenas. He was well aware of the expectations of the tourists and tour guides but didn't see why those should translate into an altered approach to his hyenas. Ironically, it was this approach that both endeared him to tourists and led tour guides to shun his business and take their customers to the other place, where disappointment lay ready and waiting. But unlike in the movies, there would be no karmic tornado to blow away the unpalatable operator. The Aboker place only got stronger.

On my seventh night at the Aboker feeding place, I arrived late and stood among the crowd, watching Mulugayta feed his hyenas. Only two hyenas were present and only one was feeding, while a crowd of tourists made a semicircle around the hyena man. I heard Saloman on the rocky outcrop calling out into the night for more hyenas to come. It was a bad night for the Aboker place; the hyenas must have found a donkey carcass out there in the farmland. As Mulugayta was feeding, he looked up and saw me among the spectators. He said something to Balay, who turned, saw me, and came over to where I was. Balay told me I had to pay more money or else leave, and began herding me away from the feeding place. I shouted out to remind Mulugayta of the deal we'd made and that he had signed a contract, but he ignored me and continued feeding. I actually didn't know whether Mulugayta was reneging on our deal or whether Tusty had pulled a scam and told Mulugayta that I was only paying to see the hyena feeding for six nights. But secretly, I didn't care. Walking in the darkness toward the Sofi feeding place, I felt glad that the decision about which hyenas to study had been made for me. I was lucky; I had plenty of time in Harar to sit around the Sofi feeding place as hyenas came and went of their own accord. I would spend no more nights in the freezing wind watching hyenas being sat on and having rocks thrown at them by low-paid workers to keep them confined. I arrived at the shrine to find Yusuf sitting by himself, smoking a cigarette, with only one young hyena lying on the ground before him. Every now and then, Yusuf threw a scrap of food to the hyena, who got up to retrieve it. I took a seat beside him and together we stared at the hyena, who was lying with her chin on her carpals. "Marcos, how is it?" he asked in Harari, using the local pronunciation for my name. I replied in Harari, *Aman inta*. It's peaceful. It's good.

# 4

## YOU HYENAS

By the time the cool December winds were blowing into Harar, I'd been visiting the Sofi feeding place almost every night for two months. Things were working out. I was being made more and more welcome by Yusuf's family and becoming increasingly familiar with the Sofi hyenas who were regular attendees at the feeding. There was Koti, the dominant female of the clan, with a massive head on an equally massive neck. Whenever the feeding hyenas turned as one and looked over to the track leading into the feeding place before clearing a space in front of Yusuf, that was the sign that Koti had arrived. She usually came in the company of her intimidating entourage, Chaltu and Diraatu, who were equally large and followed Koti's lead whenever she displaced subordinates or incited the clan to go and mark some territory.

Of the lower-ranking females, Funyamurey was easy to recognize because of the deep scar bridging her nostrils. She grumbled a lot when she missed out on food or when Yusuf compelled her to come and get food when she would have preferred to lie on the ground. Funyamurey was apparently Yusuf's favorite, possibly because of her obedience, albeit reluctant. Another low-ranking female, Guramurey, who was named for the large notch in one ear, was so stubbornly fearful of humans that she never approached to feed. Instead, Yusuf had to throw food over to where she waited in the darkness. Then there was Abba Chabsi, a placid-looking midranked female who often attended to other hyenas' cubs. Yusuf named her after an Oromo folkloric character who cared for children.

Of the males, Bouki was the highest ranked and the most confident. He also had a knack for being in the right place at the right time. Whenever someone came to dispose of a diseased livestock animal, Bouki was ready and waiting or else first on the scene. Bouki often kept company with Tukwondilli, the skinny, nervous male with a droopy nose. Tukwondilli was the only hyena who had the nerve to squeeze behind the tourists when they were sitting with their backs to the shrine, for reasons I explore in detail below. Then there were the youngsters. Ibsa was a nervous adolescent who stood in the shadows, trying to summon the nerve to approach Yusuf and take food. Once sufficiently emboldened, Ibsa dashed in and arrived in a skidding cloud of dust before snatching the food and scampering off with it. This was a little disconcerting for the tourists standing next to Yusuf, but immensely entertaining. There was Kamareya, the placid one with the big round eyes. He constantly monitored the food bucket and tried to sneak up and steal the contents whenever it was unattended. Finally, there was Baby, the little hyena with the "falling hamburger" spot pattern on her rump. Baby had a grudge against dogs and used to stand above the road and growl as dogs passed by with their owners on the way to the farmland.

At this point I should clear up a common misconception. A hyena is not a kind of dog. Take a close look at a hyena and you'll see how very undoglike they really are. They have long whiskers fanning out from their snouts and curvy back legs that are quite catlike. This is because they're more closely related to cats than to dogs and even more closely related to mongooses. Hyenas look superficially like dogs owing to their adaptation to open-country predation and their heavy reliance on smell and hearing. This is what we call convergent evolution. It's where two species arrive at similar morphologies or behaviors independently through having fitted into similar adaptive niches. But hyenas in fact belong to the suborder Feliformia, which separated from caniforms some fifty million years ago. Beyond the physical differences, hyenas' social arrangements are different from those of dogs. They live in clans of up to a hundred individuals, consisting of several matrilines of resident females and a bunch of immigrant males, all adhering to a relatively stable social hierarchy.[1] At the top of the hierarchy sits a dominant female and her subadult offspring, and below them, other females and their offspring, right down to the highest-ranking of the males, who heads a list of immigrant males, many of whom might spend just a short time in the clan's territory before moving on.[2] A newly arrived male tends to go straight to the bottom of the hierarchy and must work his way up by hanging around and letting natural male hyena attrition do the social climbing on his behalf. These social arrangements are markedly different from those of dogs, but they pale in comparison to another distinction that has profoundly affected the socioecological pathways of the two species over recent evolutionary history: hyenas are instinctively terrified of humans; dogs are not. Millions of years of competition with, and eventually persecution by, humans has ingrained in spotted hyenas an almost irresistible compulsion to run, and run very quickly, from us upright apes. This is the default reaction for hyenas from the time they are able to get up and walk. But that doesn't mean it's hardwired. Hyenas are a flexible species—which is probably why there are still tens of thousands of them running around Africa—and when it suits their needs they can overcome this instinctive fear of us.[3]

Two hyenas in the Sofi clan were classic examples of those who for different reasons could overcome their fear of humans: Tukwondilli, the skinny male with the droopy nose, and a stocky female named Dibbey. After my first few encounters with Tukwondilli, I thought him at least irrational, if not insane. Whenever he was at the feeding place, he constantly moved and dodged about as though avoiding invisible obstacles. At other times he stood motionless, simply staring straight ahead, oblivious to what was going on around him. This disregard for his immediate circumstances carried over to feeding time. Putting aside any notion of a buffer zone between himself and humans, Tukwondilli often ran through the semicircle of spectators surrounding the hyena man. For whatever reason, he was intent on being on the other side of the line, and more than a few people were visibly unnerved by this half-baked hyena brushing against their legs as he pushed past with total disregard for what is acceptable human/hyena personal space. But after night upon night of watching Tukwondilli's strange performances, I came to realize that it was not insanity that drove him to act thus; it was his infatuation with Dibbey.

Yusuf named Dibbey for her pugnacity; the name means "drum." Even as a cub, Dibbey was a menace, and she grew up to be the most overtly aggressive and violent of all the Sofi hyenas. While she was shorter and stockier than the highest-ranking females, she made up for her physical shortcomings with ferocity. She also ranked about fifth or sixth in the clan hierarchy, so there were plenty of underlings on whom she could unleash her rage. On a typical night at the feeding place, Dibbey arrived with her mane bristling and her mouth slightly open, sufficient to display a flash of white teeth contrasted against her dark snout. If lower-ranking hyenas noticed Dibbey's arrival, they quickly got out of the way as she charged at them with her head down. If lower-ranking hyenas were too absorbed in the feeding, too focused on the meat draped over the stick in Yusuf's mouth, then their first awareness of Dibbey's presence was the sensation of her teeth biting into their backs or legs. Nor was this aggression restricted to her own species; Dibbey was one of few Sofi hyenas I had seen being aggressive toward humans. Her aggression was usually directed at Yusuf's son, Abbas, or at me, when Yusuf left either of us sitting at the shrine in charge of the food bucket. In fact, it was usually the availability of food that elicited Dibbey's bursts of rage; otherwise, she could be quite placid. But there was one bee who could get into Dibbey's bonnet in any circumstances, and that was the tenacious Tukwondilli.

Granted, there were rare occasions when Dibbey showed remarkable tolerance for and even interest in Tukwondilli. Once, when Dibbey was lying on the hill, she allowed him to approach and lick her rump. On another occasion, she engaged him in a hyena greeting, which involves sniffing the other's genitals. Tukwondilli was visibly nervous and jittery at the close proximity of Dibbey's bone-crushing teeth to his vulnerable penis, and he scampered off as soon as Dibbey permitted. But these interactions were rarities. Normally, Dibbey took exception to Tukwondilli's attentions and tried to attack him at every opportunity. Yet this did not deter her persistent paramour. He was quite fixated on Dibbey and followed her everywhere she went, always arriving at the feeding place only a few moments after she did. This male hyaenid doggedness of his may be what lay behind Dibbey's apparent hostility toward Tukwondilli, fostering the strange behavior that led me to think him insane. During the feeding, he had to stay on the move in order to maintain a contraposition to that of Dibbey. He was inexorably attracted to Dibbey, but self-preservation compelled him to keep his distance. And that is where Yusuf and his chorus of human spectators figured prominently on the stage.

When Yusuf was feeding hyenas, he typically sat in the middle of the open area beside his house, which meant that there was space all around him in which hyenas could move. At about forty-five degrees to Yusuf's left, the taxi or tour vehicle shone its headlights on the scene. The tourists stood to Yusuf's left so as not to block the light, and generally formed a line between the taxi and the fence. This made the place in front the taxi and to the right of Yusuf a secure place for hyenas, and so most of them stood there. That was the direction in which Yusuf offered most of the food; that was the place where the hyenas could keep their distance from the spectators. On Yusuf's far right was the path to the hill and darkness, where the more skittish hyenas watched the proceedings. Meanwhile, in the space behind Yusuf's back, with spectators to one side, there stood Tukwondilli, nervously eyeing Dibbey.

This was how Yusuf was drawn into relations between the star-crossed Tukwondilli and the very cross Dibbey. Yusuf was like the bush around which two comic characters circled, the monkey and the weasel all around the hyena man. A bush who was very much aware of his centrality to the proceedings, however, because Yusuf often called out Tukwondilli's name as he threw a scrap of meat over his shoulder. This was of course anathema to Dibbey, who charged around to the rear of the hyena man, intent on attacking Tukwondilli, who had to quickly snap up the food and run past the spectators to the front.[4] Then, upon Dibbey's return to the front, Tukwondilli scampered around to the rear and the dance continued. Meanwhile, Yusuf reprimanded Dibbey for chasing Tukwondilli and ordered her back. Often, he threw a piece of food over Dibbey's back, whereupon she turned and chased away any hyenas attracted to the flying meat scrap. Then, unseen by Dibbey, Yusuf quickly flicked a piece of food to the nervous hyena standing to his rear.

Intriguingly, Tukwondilli's fixation on Dibbey and the imperative to keep a safe distance outweighed his fear of people. This is where the line of tourists at the feeding became central to the show. On occasions when Dibbey's chases became persistent, Tukwondilli headed for the safety of the shrine, uttering barely audible alarm rattles.[5] If there was a line of tourists in front, Tukwondilli simply charged through the line, leaving Dibbey on the other side of the impenetrable human barrier. If it was a quiet night and there was no line of tourists, there was always someone—frequently myself—sitting on the high step in front of the shrine. Tukwondilli found that on these occasions he could retreat to the space between the shrine and Yusuf's fence, adjacent to where I was sitting. This was too close to me for Dibbey's comfort and she always pulled up short, staring hard at Tukwondilli before heeding Yusuf's calls to desist and return to the feeding. Yusuf was also aware of Tukwondilli's reliance on the shrine for protection, and he occasionally employed it in his feeding show. Whenever he had the inclination, he stood up, called to Tukwondilli, and carried the bucket of meat scraps over to the step of the shrine, where he took a seat among the tourists. Then, while Dibbey watched from in front, boiling with impotent rage, Yusuf draped a piece of food over the stick in his mouth while Tukwondilli squeezed behind the seated people's backs to get to Yusuf and take the offering. Needless to say, this could be pretty exciting for some of the tourists.

Tukwondilli's darting around and through crowds of people raised some questions for me about his relations with the world other than Dibbey. When I was sitting at the shrine, watching Tukwondilli circling around Yusuf, breaking the line of people, or feeling him pressing against my back as he squeezed between me and the shrine, I often wondered, what on earth does he think we are? A normal hyena would shy away from a line of people, as Dibbey always did. No matter how hungry or fixated on food they were, the hyenas were always very aware of the human presence and made certain not to come too close. Yet for Tukwondilli, we humans seemed mere obstacles to be used in keeping Dibbey at a safe distance. Even Yusuf drew very little of his attention; Tukwondilli took the pieces of food almost as a distraction from his focused attention on Dibbey. It was as though, in Tukwondilli's eyes, humans and other hyenas simply blended into the landscape as objects around which he and Dibbey danced.

I tested this theory one night when I saw Tukwondilli lying in the darkness on the hill behind the feeding place. Considering the audacity with which he pushed past people in a feeding context, I wondered if he would remain on the ground while I went and sat down next to him; I wondered if I would retain my object status in circumstances in which he was not so distracted. So I walked slowly but directly toward him. On hearing me approach, Tukwondilli lifted his head and looked directly at me. Then, when I was about four meters away, he quickly stood up and hustled off to a safe distance. The moment called for a fanfare, but I contented myself with a chorus of crickets.

So how is it that Tukwondilli could push against my back in one set of circumstances, as though I was an obstacle, and yet actively maintain flight distance in another? How was it possible for me to be transformed from a something to a someone in the eyes of a hyena, and then back again? Obviously, context is an integral part of the answer, but only a part. The Austrian-born philosopher Martin Buber provides a useful theoretical framework here, whereby he conceives of different kinds of relations based on ways of attending to the other. Buber was interested in the coming together of selves and how those selves' relations stand at the foundation of personhood. For Buber, the sphere of relation in which persons are actualized is that in which they address the other as capitalized You. He refers to this as the I-You relation.[6] Counterposed to this is the I-It relation, in which the other is a thing with which a person can also enter into relations but without the resultant actualization of self. In fact, according to Buber, the person who enters into relation with You does not stand in relation to anything at all; they simply stand in relation. The reason is that when the You becomes a something, it can no longer be a You; it becomes an It. The relevance for me lies in the way that Buber holds the I-You relation to be fleeting. As a person moves across relations, "every You will enter into thinghood again and again. Every thing will appear as a You." Moreover, there can be no tripartite relations; the self can only attend to one You at a time. In order to enter into relations with another You, the previous You reverts to being an It. In the case of Tukwondilli, we can imagine that his You-saying toward Dibbey was so profound and protracted that he had very little opportunity, or inclination, to move on and engage with another You. If I was to succeed in engaging his attention, I had to choose a time and place in which he was open to an I-You encounter with another being.

Still, I had reservations about the extent to which Tukwondilli considered me a person, the extent to which he recognized my intentionality. Although he had indeed looked directly at me and moved away, I could argue a case that he considered me nothing but a self-propelled zombielike thing that would step on him unless he got up and moved away. In that respect, I was not unlike a minibus maneuvering in the feeding area; vehicles reversing blindly toward Tukwondilli usually (but not always) drew his attention, and he sidestepped to get out of the way. But Tukwondilli made a difficult test case for hyena/human intersubjectivity. At his best, he was still a little unhinged and very difficult to read. It was another hyena who would not only engage with me directly but also recognize me as a capitalized You. And this was because our intentions converged on a burlap sack full of shredded paper.

Probably the most audacious of the Sofi hyenas and the one most "into" the human world was the little hyena named Baby. One night I sat down with Yusuf in front of the shrine and, with sports journalist Mohammed Sayeed translating, Yusuf told me Baby's story. "Her mother was Jambo," he began. "This Jambo, she was sick the time when she brought Baby to the feeding place. First time. After that, Jambo, she disappeared. She must have died. So I took Baby to the forest to show her where the other hyenas live and since that time she comes every night. Baby is always first to come." How did you choose her name? "Oh, it was a farenjo gave her the name. One night some farenjo tourists came to feed and a woman was saying, 'Look at the baby! She's just a baby!' So I called her Baby. She knows that's her name." How long has she been coming to feed here? "Maybe three months." Considering her size at the time, she must have been pretty small when her mother died. I asked, How big was she when she first came here? Yusuf's son Abbas, who was sitting beside me, leaned over and held his hand about six inches above the ground.

Yusuf was quite fond of Baby. When he was feeding the hyenas, he called Baby's name a lot more than the others and made sure he threw food over to where she waited behind a wall of big female rumps blocking her access to the hyena man. Then, when she began standing behind Yusuf, he made sure to turn around every now and again to pass her a piece of food. In time, Yusuf became so familiar with Baby's habit of standing behind him, and she so confident in his knowledge of her presence, that something magical began to happen: without even looking to see if Baby was there, Yusuf pointed the feeding stick backward over his shoulder and called Baby's name, whereupon she reached up and gently took the meat off the stick. Perhaps this is what Lim Ka Ea was looking for: the harmonious middle ground where human and hyena meet on their own terms and arrive at a mutual understanding. Yusuf was attending to what Baby was doing. He was treating her as an individual, looking into her hyena world and understanding what she was doing there behind his back. He could have withheld the food and compelled her to go around to the front, like the other hyenas, which certainly made more sense in terms of the feeding show. Instead, he took a step toward the hyena world, toward the middle ground of intersubjectivity that requires an attempt to understand the Other. But was she seeing into Yusuf's world?

The thing I found most endearing about Baby was how she wore her heart on her sleeve. In this respect, she was like a comic actor in a silent movie. When a big female hyena's snout collided with her own during the feeding, making an audible "thunk," Baby showed her hurt to the world. She turned away, snorted, licked her nose, and, with her head and shoulders slumped forward, shuffled over to the dumpster, where she lay down with her chin on her carpals and sniffed. When an Ethiopian tourist standing beside Yusuf panicked and dropped the stick in front of Baby, she instinctively followed the stick toward the ground, but she threw her head up when the man's friends burst out laughing, and she swiveled her ears around while looking wide-eyed from Yusuf to the crowd and back again. When she arrived at the feeding from the wrong direction and found a wall of tourists between herself and the hyena man, she almost scratched her head as she summed up the situation. Then revelation, and she headed into Yusuf's compound, crossed the yard, and emerged through a hole on the other side of the tourists. Surprised, confused, hurt, dejected: Baby let the whole world know.

The first few nights that I arrived at the Sofi feeding place, Baby stood up in anticipation that I was a tourist. She looked from me to Yusuf and then back to me, but once she understood that there would be no food associated with this farenji, she lay down again, plonked her chin on her extended carpals, and exhaled. Her disappointment was not only palpable; it was hard not to take personally. Then, after about three weeks of nightly visits to the Sofi feeding place, Baby showed barely any interest in me. On my arrival, she didn't even bother to get up or look at me; she just lay on the ground, facing forward, and followed my footsteps with her ears. When I took a seat at the shrine, she lifted her eyes to look at me but only for a moment, before turning them away to stare straight ahead again. She was waiting for tourists and food, and I was just a spectator. But as far as my research was concerned, this was progress. As she became less interested in me, she also became less concerned about me as a potential threat, and the buffer zone that she maintained between us, small as it was, became even smaller with each uneventful night. I anticipated that before long I'd be able to follow Baby if she ventured into the Old Town, and make some meaningful observations of an urban hyena's travels in Harar's lanes.

With each passing night, Baby's trust of me increased. When Yusuf emptied a bag of scraps onto the ground in front of the hyenas, Baby emerged running from the feeding frenzy with a cow's tail in her mouth and a bigger hyena in pursuit. Head held high, ears folded back, and the tail trailing between her legs, Baby bolted toward where I was sitting beside the shrine. Before I even had a chance to move, she ran past me within inches, while the pursuer pulled up short, snorted, and with a swish of her tail returned to the feeding. On another occasion, when I was walking from the hill behind the feeding place to the road, I had to pass through a group of hyenas lying in the middle of the feeding area. Although I tried not to disturb them, they got up and shuffled away in deference to their requirement for a buffer zone. All except for Baby, who remained lying on the ground as though she hadn't even realized I was there. She just followed me with her ears as I passed. Baby became ever more comfortable with my close proximity, and I consistently pushed at the edges of her buffer zone. But for all I knew, this trust did not necessarily reflect any kind of recognition of my subjectivity. She may well have decided that I was a useless and apparently harmless self-propelled object in an object world. She only revealed her hand, or drew a new one, when I took a step into her relational sphere.

Baby had a predilection for human-made objects. More than any other Sofi hyena, she was mad keen on Harar's urban detritus. She often arrived at the feeding place with a rubber sandal in her mouth, or spent her time chewing on plastic bottles. Once she even dragged a length of PVC plumbing pipe across the road and dropped it next to the shrine without even bothering to chew on it. If something was unattended and human-made, then Baby was interested, and she either carried it off to chew, walked around with it in her mouth, or, if it was too big to carry, chewed on it in situ. One night I found Baby sitting beside Yusuf's compound with a burlap sack full of shredded paper. Given her capacity to rip the sack to pieces, she was quite gentle with it and just chewed on one of the corners. Without thinking things through, I went over and sat down on the ground within reach of the sack. Baby stopped chewing and stared at me as I sat down, but she resumed chewing once I was settled. Without a word, I reached across and gently grasped the corner of the sack. Baby let go of the sack, lifted her head, and opened her eyes wide as I slowly pulled the sack toward me. She sat up, looking perplexed, but apparently resolved to get the sack back. She edged closer, leaned over, took a bite of the corner, and dragged the sack back toward where she'd been sitting, where she resumed chewing. Not wanting to leave it at that, I reached across once more to grab the corner of the bag. But Baby saw my move, and before I could get hold of the corner she pulled the bag away from me, stood, and raced off with it across the feeding place. Naturally, I gave chase.

The other hyenas scattered as the two of us careered across the open space in front of the shrine. Baby bolted across the little bridge over the narrow drainage canal and onto a patch of grass beside the road. I anticipated that she would head for the farmland, so I jumped the canal and headed her off. Seeing my move, she turned about and ran toward the road, shreds of paper flying in her wake. I ran after her into the darkness and saw her running along the line of Yusuf's fence to a point at which she disappeared. I arrived at the spot out of breath and found I'd lost her. Then I heard the sound of shredded paper being crunched and saw, through a small opening in the thorn fence surrounding Yusuf's house, Baby with her prize. All I could do was watch while she chewed victoriously on her burlap sack behind a wall of sticks and thorns.

Walking home that night, I was excited about that little tug-of-war and how it was going to afford me progress in my research. I'd been reading the work of the primatologist Barbara Smuts and her views on habituation.[7] For Smuts, the habituation of baboons was more a matter of entry into the baboons' social spheres than it was the researcher's making herself invisible to the animals. Becoming invisible was not an option, as the baboons demanded that the human interlocutor engage with them on their own terms. Once Smuts recognized this, she was able to blend into baboon society and gain profound insights. I was focused on a goal and determined to achieve it. I needed to somehow habituate some of the Sofi hyenas in order to follow them around the Old Town at night and collect meaningful data. Without the data, I'd have nothing to write about and nothing to support the first nothing; I'd be sunk. In the moment when I reached out to grab the corner of the burlap sack, I was guessing that engaging with Baby on hyena terms, competing over the sack, would admit me into Baby's social sphere and set me well on the way to habituation and hyena followings, and that a shiny new doctorate would follow soon thereafter. Big mistake.

After the night of the burlap sack, Baby certainly recognized me as an intentional subject. I had shown myself to be a creature capable of the same interests as herself, a fellow fancier of burlap sacks who had the same predilections that she did. I was intentional, I was self-motivated, and my intentions could, in the case of a burlap sack, be gauged. But the upshot of this was I could no longer be trusted; Baby began avoiding me. Whenever I went to sit near her at the shrine, she got up and left. When she emerged from a mass of hyenas with a prize, she changed direction and ran away from me. She became far more wary of me than before, and the space she kept between us trebled. In terms of my research aims, this was an absolute disaster, and I felt like I'd been set back to the square just before square one. Ironically, Baby's recognition of my subjectivity and my intentionality turned out to be detrimental to what I was trying to achieve in Harar. The hyena who had been showing the most trust in me became the least likely to allow me to follow her and make observations. Moreover, I couldn't see how to repair what I'd done except to try and be innocuous for God knew how long. I'd set myself back months because of a harebrained idea to steal a burlap sack full of shredded paper.

So a hyena had come to recognize my intentionality and subjectivity. Big deal. In terms of my research aims, I was frustrated, and the night of the burlap sack was a disaster. Habituation was turning out to be far harder than I had anticipated. Most of the hyenas were too skittish to ever allow me to come close in a non-feeding context, and where I had made initial progress with Baby, I had blown it. Identification of individual hyenas was nearly impossible. Most never allowed me to get close enough to take flash photos for a database, and when they were feeding in the taxi headlights, they all faced toward Yusuf, giving me no chance to photograph their left sides. My attempts at systematic data collection were also floundering. I'd thought of asking the police to record the locations of their nightly encounters with hyenas on maps that I provided, but the few policemen who were interested in the idea couldn't read maps. I was struggling with the language. I'd learned a bunch of words and phrases, but I was stumped when it came to linking them together in any comprehensible fashion. I was frustrated in my attempts to interview people. When I arranged a time, the respondents either failed to show up or, if I visited them, they complained that they were sick or upset and asked if we could do it another time. This was compounded by my own health problems. Between giardia, flu, bronchitis, and unexplainable dizziness, I was spending as much time at home in bed as I was out collecting data. And bearing down on me like a train, the night of Ashura was approaching fast.

# 5

# THE LEGEND OF ASHURA

A long, long time ago, there was a terrible famine in Harar. The rains failed, the crops withered, livestock lay dead on parched soil, and people lay starving in the streets. At the same time, the hyenas had begun attacking people. Every night they came to the town and took the weak and dying, dragging them off to be eaten. The people of the town became frustrated and protested in the streets, marching with stones on their heads to represent the weight of the world that was oppressing them. They demanded that something be done. The town's saints responded, and at their weekly meeting on Mount Hakim they decided that the hyenas should be fed a multigrain porridge. They passed the message to the people in the town, and the porridge was prepared and offered to the hyenas, who ate it happily. A truce was made. Since that time, on the eve of the tenth day of Muharram in the Islamic calendar, porridge known as *shure* is prepared and given to the hyenas around Harar. This reinforces the truce and ensures that no people will be attacked in the following year.

On the eve of Ashura, people gather at various shrines outside Harar and pray while the porridge is prepared.[1] Once ready, the porridge is taken outside, poured into stone bowls known as *waraba gabata*, or "hyena bowls," and covered with lashings of butter. Then the hyenas arrive. The leader of the hyenas, the *waraba amir*, is the first to approach. He is completely white, dignified, and very handsome. He gazes into the butter and, if he sees his reflection, begins to eat. Sometimes the porridge is not salty enough and more salt must be added until the waraba amir is satisfied. Once he is happy with the taste, he signals to the other hyenas to come and join him. They all converge on the porridge and eat peacefully. The amount that the hyenas eat is important. After the hyenas have finished, the caretaker of the shrine, the *murid*, inspects the leftovers. If the hyenas have eaten every bit of porridge, then famine will come in the following year; if they eat only a little, then misfortune and war. If they eat most of the porridge and leave just a bit, then it will be a prosperous year for the townspeople.

That is essentially the story of Ashura. There are numerous iterations, in which the details vary. In one telling, the solution to the ongoing attacks is found when some hyenas intending to attack a woman choose instead to eat the porridge that she is making; in another, a boy with a pure heart dreams up the solution. Abadir figures in some versions as the one who arrives at the solution and directs the townspeople to make the porridge. Emir Nur figures in another.[2] But the underlying theme is essentially the same: famine in the town, hyenas attacking people, solution found, porridge fed to hyenas every year thereafter, handsome white hyena king, truce maintained, fortunes of the coming year foretold.

I knew about Ashura and the porridge feeding before I came to Harar. My Harari-Australian contact had told me about it, there are accounts of the ceremony in Camilla Gibb's dissertation, and it is the subject of an article by Harari scholar Ahmed Zekaria.[3] But firsthand accounts are hard to come by. Gibb describes the story and the events of the celebration, but she was unable to see them herself because of a curfew at the time of her fieldwork. Zekaria's account is more promising but unspecific. He simply says that filming was done at the shrine of Aw Aboker and that he was looking forward to the edited result as a step forward in ethnographic filmmaking. Still, I was hopeful that some kind of porridge feeding might take place at one of the shrines, so I'd added "documenting and filming the events of Ashura" to my list of things to do in Harar.

In the lead-up to Ashura, I told my sports journalist friend Mohammed Sayeed about my plans to film the hyena porridge feeding. He responded, "Oh, Mr. Kangaroo, this is going to be a difficult program. You're lucky you have Mohammed Sayeed as a friend to help you." He often referred to himself in the third person like that. Mohammed took me to meet Abdella Sherif, the operator of Sherif Museum in Jugol. A dynamic sort of Harari, Abdella built his museum out of nothing. By going door to door asking for old artifacts, which he displayed in his house, he'd built up his collection and taken over the old palace of Ras Makonnen, which he was in the process of restoring.[4] Without much dependence on the Harari government, he continually sought funds and equipment from overseas in pursuit of his mission to preserve Harar's heritage. We found him in a burcha house off to one side of his museum, where he'd set up a device for photographing old Harari manuscripts page by page. He was a big man with a loud raspy voice and energy to burn. Shaking my hand vigorously, he welcomed me inside and asked about my research. I outlined my plans and asked if he'd ever seen the porridge feeding at Ashura. "We tried to make a film of it," he said, sounding like he was gargling gravel. "Three years running we went to Aw Aboker with a camera, and every year nothing. No hyenas, nothing." It didn't sound promising. "You know the cameraman with the culture bureau, Mohammed?" I had seen him around. He had a magnificent head of hair. "Yes, him. He tried for fourteen years to film the hyenas. Different places, Aw Hakim, Aw Aboker. But nothing." Before I left, Abdella gave me a DVD entitled "Dermesheh." It had a picture of a hyena on the cover. "Have a look at this, it will help you with your research." I thanked him and left feeling a little bit pessimistic about filming hyenas eating porridge.

Mohammed Sayeed wanted me to meet someone else at the museum, so he took me to an office beside the main entrance. Before we entered, he whispered to me, "This man can help you, but be careful. He's a waraba with money." Inside sat a man at a computer, wearing a Muslim cap and gown. The man looked up at me and then looked to Mohammed Sayeed with an unspoken question about this farenjo who'd been brought into his office. Mohammed Sayeed introduced him to me as Ali and, reading the look on Ali's face, explained what I was doing in Harar. "Excellent," said Ali, smiling with mottled teeth. "Tell me, do you chew khat?" Just a little, I admitted. "Then we shall have burcha together and I will tell you all about the warabas. Do you receive funding from your university?" I replied in the affirmative. "Then we shall be able to find a way that we can both benefit." I explained that my funding was limited and that when it ran out I would be digging into my own pocket to continue my research; in other words, there was no money tree growing in my garden. "No matter. If Allah wills it, there is always a way." We arranged a time for burcha and said our good-byes.

Back in my apartment, I played the DVD. It was a nicely put together documentary about Harar's hyenas. There was footage of hyena feeding, hyena legends, and numerous interviews with locals and some academics about the importance of the hyenas to Harar's heritage. It was in fact a sort of filmic version of what I was trying to achieve with my ethnographic research on Harar's human population. Toward the end of the video, the subject of the Ashura ceremony was raised. The narrator described the history of Ashura, interspersed with comments from experts and locals. Then, accompanied by a dramatic score, footage of hyenas eating porridge. I studied the footage carefully and my heart sank as I made out the details. The porridge was served up in a single tin dish, and in the background, there was Yusuf's fence. The filmmakers had contrived an Ashura ceremony at the Sofi feeding place.

By this time I was very skeptical about the porridge feeding. The only eye-witness account I'd found was from the French-Harari woman, Yasmin, who said she'd seen it as a child. But her description of the events was closer to the local legends than to any kind of porridge feeding I imagined took place. Everyone else who described the Ashura ceremony finished with the disclaimer that they had never personally seen the porridge feeding. And having seen how skittish the nightly provisioned Sofi hyenas were, I couldn't imagine that hyenas out in the countryside where the shrines were located would simply march up and start eating porridge in front of a bunch of human onlookers. It was all becoming too far-fetched, so I wasn't particularly motivated to go to the trouble of traveling to one of the shrines in the vain hope of documenting hyenas being fed porridge, when it appeared that the chances of doing so were extremely slim. Sure, it was meaningful that every Harari knew the legend of Ashura, but in terms of an ongoing practice that I could document, I wrote it off. Then my phone rang.

The man on the line spoke excellent English in a very cheery voice. His name was Nureddin and he'd been given my number by Mohammed Sayeed. Nureddin was marrying an Australian-Harari woman and was in the process of applying for a visa, so he wanted me to get to know him and his fiancée and vouch for their marriage at some point. I had no problem with that, so I agreed to meet with him at the main market. Once there, I called Nureddin and he came to meet me. He was a sturdy-looking man though by no means tall, with an Arabic nose, shaved head, and a hint of a moustache. We went to his house, where I was introduced to his fiancée, Hasanat, who was making coffee, and a group of Nureddin's friends, who were scattered about the living room chewing khat. The air was thick with cigarette smoke and the smell of incense and the floor was littered with *garaba*—the khat leaves that are too old and bitter to chew. Nureddin indicated that I should sit on one of the platforms and passed me a bunch of khat. One of the Harari men asked me what I was doing in Harar. When I told him, he translated to another who spoke little English. The man spoke at length in Harari and Nureddin translated. "He says that during Ashura, the hyenas gather at the shrine at dusk. They know the time of year and come every time. The leader is completely white and very handsome. He comes with four others and tastes the porridge before the others can eat." I asked if the man had seen this himself and the reply came back: "No." Nureddin added, "Maybe this is true, maybe this is myth. Nobody knows."

That night, I went to burcha with Ali, the clerk from Sherif Museum. He had a modest house in the Argobberi quarter, where I found him chewing khat and typing away on a laptop. He began by asking how my research was progressing. As I detailed what I'd been doing and some of my findings, he dragged on a cigarette, staring off to one side and grunting occasionally. I had the impression that he wasn't listening but was waiting for an opportunity to say something, so I stopped midsentence and stuffed a few khat leaves in my mouth. Ali turned, gave me a protracted meaningful look, and said, "I have found some information for you about Ashura." He paused and studied my face. "If you are looking for a place where there is guaranteed to be hyenas, then you should go to Aw Nugus. This is a shrine near to Kombolcha. The hyenas come every year to the place on the night of Ashura, I am certain of this. The people make the celebration and the shure is prepared by the women. They put it like this on the *gabata*, the stone, and there is three." He arranged three espresso glasses to form a triangle. "The hyenas come to this place every year without fail." Sounding a little too skeptical, I asked bluntly, "So who told you about all this?" Ali looked shocked and insulted. "The elders, of course."

With less than a week left before Ashura, Nureddin invited me to coffee at Hakim Café just outside the Old Town. I was always keen to go to this café, specifically because of a beautiful Oromo woman named Tigist who worked there. Seven months later we would be married, but I never imagined this at the time. Nureddin asked how my research was going. I told him about the progress I'd made, or the lack thereof, and of my skepticism about the Ashura festival. He agreed, saying that it sounded more like myth than reality, but why not find out for certain? "If it is a myth, then so be it, but if not, then you should go and make a film of this Ashura ceremony." From that point on, Nureddin made it his personal responsibility to make sure that I went to a shrine to try and film the porridge feeding on Ashura.

Our first step was to visit the director of the Tourism, Sport, and Culture Bureau and ask for assistance. I was surprised by the deference that the people in the office showed to Nureddin and by how readily we were admitted to the director's office. Before we went in, Nureddin confided in me, "We are old rivals, the director and myself. We went to university in Addis together and then joined opposing political parties. But I will impress upon him the importance of your research and demand his help. If he is serving the interests of Harar, then he cannot refuse." As we entered the office, the director stood to greet us. He was a bear of a man, with a full beard and eyes like dark stones. He shook hands with Nureddin and they talked in Harari for some time. Then he looked to me and said, "I was trying and trying to work out why Nureddin is helping you, and now I see. It is because he wants to go to Australia." I replied in my broken Harari, "Or else he wants to be a farenjatali." They both laughed the kind of laugh that quickly fades. In the discussion that followed, Nureddin demanded a vehicle that would take us to one of the shrines on the night of Ashura, expounding on the importance of my research to Harari cultural heritage. He also asked for a letter from the director explaining what I was doing in Harar and stating that no one should interfere with my research. Nureddin took me into the reception area and stood over the typist while she typed, then rushed off to another office to get an official stamp.[5] The director's receptionist commented that he was "like a lion." I said it was more polite to say he was a waraba. Within half an hour, we left with the promise of a vehicle and a document that would in time become my key to the whole regional state. I'd never seen Ethiopian bureaucracy move so fast.

Two days after our meeting at the Tourism, Sport, and Culture Bureau, Nureddin called again and asked me to meet him at the Bureau of Information. On arriving, I found him sitting with a man wearing a coloyta, a cap favored by Muslim men. Nureddin introduced him as Abdullahi, the murid of Aw Nugus and the direct descendant of the man to whom the shrine was dedicated. He was tall, with Arabic features and benevolent eyes. Not confident in his English, he spoke through Nureddin about the Ashura celebration. "The important thing is to control the children so that the hyenas do not run away. If you let the children do whatever they want, then the hyenas won't come. To start with, everybody prays until the hyenas come. At this time they are very afraid of the people. Then, after the hyenas eat the porridge, a goat and an ox are slaughtered and the leftover pieces go to the hyenas. After this they become very brave." This sounded a lot more plausible than previous accounts I'd heard. I asked if there were years when the hyenas failed to come. "No. They come every year. This thing is certain." So he'd seen it himself? Abdullahi already understood my question and preempted Nureddin's translation. "Yes, I have."

There it was: an eyewitness account of hyenas eating porridge at Ashura. Not only that, but it was the murid of the shrine where the hyenas were fed, the same shrine that Ali had mentioned. Things were looking pretty promising; at least we had a strong lead and a specific shrine to visit. But not every wind was blowing in our favor. Nureddin's old rival called and said that a vehicle would not be available to take us to the shrine. It was a setback, but Nureddin was unfazed. He said he'd call the manager of the local government TV station and impress upon him the importance of what we were doing so as to borrow a car. With his flair for melodramatic speech and his clear pronunciation, he declared, "No obstacle will stand between us and the festival of Ashura. We will determine whether this is a myth or reality!" Like me, he had lingering doubts about whether we would ever see hyenas eating porridge.

On the morning before Ashura I got up early and charged every battery I had. Even if there was only a miniscule chance of filming a hyena eating porridge at

Aw Nugus, I wasn't going to blow it with a flat battery. Besides, I wanted to film all of the goings-on at the shrine, the prayers, the porridge making, and so forth. I called Nureddin and found him sounding sombre. "Our plan has been sabotaged. The manager of Harari Television decided to take the car to Dire Dawa for a wedding and left us with nothing. Meanwhile, he is sending a crew in another vehicle to Aw Nugus to film the Ashura celebrations, but they say there is no room for us. They are doing this because of who I am." I asked about our options. Could we take a minibus? Rent a car? "Don't worry, I am making some calls to find another vehicle. Meanwhile, you should come here after lunch for burcha." Uh oh. I'd become very familiar with what happens after burcha in Harar: nothing. People go quiet and continue chewing khat until it's time to go home. This is why the brief lunchtime burcha is called "lightning." It's not a real burcha because people go to work afterward.

I went to Nureddin's house feeling edgy and powerless. I wished that somehow I was able to get things happening by myself, but I simply didn't have the contacts. Nureddin was the best chance I had. The only chance. Inside his house, a few men were chewing khat and smoking cigarettes. Hasanat was preparing lunch, and Nureddin was in his usual place in the living room with a bunch of khat, talking animatedly on the phone. He nodded when I entered and put up a hand to greet me while continuing his conversation. I sat and accepted some khat. Nureddin wound up the conversation, hung up the phone, turned to me, and said, "*Abshir*, relax. We have a vehicle to take us to Aw Nugus. So let us have something to eat, finish burcha, and then we shall go." It sounded like we'd be on our way in minutes, but it was a frustrating three hours before we left the house. We walked to Feres Magala, where we found a friend of Nureddin's waiting for us in an old, beige-colored Land Cruiser. As far as I was concerned, it was the shiniest car in Ethiopia.

Driving through the hills to Aw Nugus, you realize just how much the local economy depends on khat. It's everywhere, row upon row of the emerald-green bushes, interspersed with the odd mango or coffee tree. Wherever the soil can be turned and farmers can cling to the side of a hill, they plant khat. We stopped at the town of Kombolcha to buy khat from the roadside market—kind of like a drive-through khat outlet—and then took a minor road into the hills. Another ten kilometers and we found the hillock at the top of which was the shrine of Aw Nugus. The reporter and cameraman from Harari TV arrived just after we did, and I noticed the considerable space available in the back of their car.

The shrine was a complex of old stone buildings on the hilltop. Looking back in the direction from which we'd come, I could see rolling hills covered in khat. In the other direction, west, the hills dropped rapidly away to the arid Rift Valley, behind which the setting sun was a deep orange. We were right at the edge of the Harar Plateau, at the limits of the old Harari territory, and I wondered about the man, Aw Nugus, who had come here so many hundreds of years ago to teach Islam to the locals. I followed our party through the compound gate and into a central courtyard, around which there was a collection of stone buildings. Abdullahi, the murid, greeted us and showed us around. There was the men's room, the women's room, and the place where they made the porridge. Then he took us outside to a place below the wall of the complex where there were some large stone slabs set in the ground. As the murid talked, Nureddin whispered, "This is where they put out the porridge for the warabas. Apparently, it previously sat above the ground level but it has been inundated by the earth." We were shown a place on the wall from which I could film, and then we were taken to the men's room.[6] We tiptoed across bags, legs, and bunches of khat to the far wall, where Nureddin established a sort of base camp. I deposited my gear and went out into the compound with another member of our group, Khalid, who smoked cigarettes like a French film star. Khalid was proficient in Oromo, so he was to be my guide and translator as we explored the goings-on at the shrine.

Across the compound in the women's room, the lighting was minimal, but things were more upbeat. The women were clapping their hands and singing while children played around the room. Two sheep stood in one corner slurping up discarded khat leaves. (Sheep and goats absolutely love khat, but it completely trashes their road sense.) Behind the men's room, under a veranda, the atmosphere was positively celebratory. Several Oromo women were fussing over a huge pot of porridge that two of them were stirring with long wooden poles. With characteristic Oromo hissing as they exhaled, the two women put tremendous effort into stirring the porridge, while another added more salt or flour. Khalid translated that they were not allowed to taste the porridge; the hyenas had to taste it first. As I filmed, the women began singing in Oromo.

> You headstrong protagonists will make peace
> When you return for this supper.
> These good women will prepare you this feast
> If only you will make peace.

Back in the men's room, the mass of shoes inside the threshold had grown larger as more and more men arrived and joined the prayers. Beyond the shoes, the floor mats were being increasingly covered with khat leaves. There were no chairs; everyone was sitting on the floor, and those around the perimeter of the room rested their backs against the walls. The Quran readings wound up with eighty-nine repetitions of the phrase *Lah Ai'lah, Ai la'Allah.* "There is no God but God." The chant was hypnotic and relaxing, and everybody took a collective breath on its completion. It was not unlike the end of a yoga session. The murid came over and said a few words in Harari to Nureddin. Khalid translated as the murid spoke to me. "Our hyena is unique. He is peaceful with us because of this Ashura. You will see. He will come." As the men chewed khat and talked among themselves, a boy went around the room with a strong-smelling imported perfume, a few drops of which he dabbed on people's necks. The cameraman from Harari TV was circulating around the room, pointing his camera directly in the men's faces, as is the style of Ethiopian news videography.

Just as I was wondering if I should have a look at what was going on outside the men's room, a man stuck his head into the room and called to Khalid, who quickly stood and summoned me to follow him outside. At the same time, the cameraman and journalist grabbed their gear and hurried after us into the semi-darkness. We went around to the veranda where the porridge was being prepared and found a cluster of large wooden bowls, each with a doughnut-shaped circle of porridge inside. Three of the women had each taken up a bowl, which they placed on their heads before gliding gracefully out of the compound. They weren't waiting for us or the news crew; they had an agenda of their own and were sticking to it. We hurried after them with flashlights and equipment, like paparazzi.

Outside the compound, on the stone slabs that I'd been shown earlier, the porridge was turned out of the bowls to create a triangle. I immediately remembered the three glasses that Ali had laid out on his living room floor. This was it, the legend of Ashura playing out before my eyes! Unable to get clear footage of what was happening in among the crowd of people, I remembered the vantage point on the wall I'd been shown by the murid. I raced back around to the compound and up the steps to the high ground where the shrine proper stood. As I reached the top of the steps, I encountered three armed soldiers standing guard over all of the local children, who had been herded into a square as if they faced imminent execution. The chances of hyenas coming to feed seemed to be increasing with every moment. I rounded the bend and chose a suitable place on the wall, at an oblique angle and a considerable distance from the porridge, so as not to scare away any hyenas, should any materialize. I started my camera and pointed it downward to film the crowd of people with three mounds of porridge in their center, the whole scene illuminated by a single kerosene lamp. The TV cameraman was weaving his way through the people, filming faces, porridge, and more faces. Then an elderly Harari woman took a tin of liquefied butter and poured some into the center of each mound. Satisfied with the arrangement, she led the people away from the porridge. As the light of the kerosene lamp disappeared with the people going back into the compound, I switched off my camera, making sure to leave it set to night-shot mode. Nureddin arrived at the wall and bade me good luck. "We shall go inside now and wait. I hope you will be successful." There were only a few of us at the wall: a soldier, the news crew, the murid, a couple of elders, and a Harari girl with a digital camera. I guessed that the girl must have been the daughter of someone important; otherwise, she would have been made to go inside with everybody else. Besides, only the wealthy in Harar had digital cameras at the time. In near total darkness, we all sat quietly. I wondered how long it would be before any hyenas came to try the porridge—if they came at all. I adjusted my position and shifted my awareness to the discomfort of the wall on which I was going to be sitting for some considerable time.

While I shifted my rear off one particularly uncomfortable stone, I noticed movement in the shrubs below the wall; a large hyena emerged and began smelling the porridge. In the time it takes a hyena's nostrils to flare, I went from uncomfortable to gobsmacked to panicked and began fumbling with my camera controls to get the thing to start up. The TV reporter, who had positioned himself on the wall directly over the mounds of porridge, switched on his flashlight and shone it at the hyena, who looked up nervously. I urged my camera on, expecting the hyena to run off at any second. She was obviously very wary of the people but at the same time drawn to the porridge. The LCD viewfinder lit up with the eerie green glow of the night-shot setting, and I pointed the camera at the hyena, who by this time was slurping butter from the mounds of porridge. I barely noticed the breath on the back of my neck as the soldier standing behind me drew close and looked over my shoulder at the little screen. I was amazed. I forgot about the cold and the discomfort of the stone wall and became absorbed with my own absorption. I thought to myself, this is truly the most amazing thing I have ever seen. That hyena was waiting in the shrubs for the people to leave the porridge alone and had emerged almost immediately to eat.

But that hyena was also about to run away. Her tail was wedged firmly between her legs, ears cupped forward toward the people on the wall, body arched backward so as to make every last millimeter of space between herself and the people count. She kept looking up at the TV reporter's flashlight and jerking backward before holding her flight. Meanwhile, the girl with the digital camera was taking flash photos from directly in front, and the hyena kept stopping to look up at the flash. I hissed at the reporter to switch off his flashlight, but he ignored me. My infrared searchlight was useless to him; it illuminated nothing for his standard video camera. I hissed at the girl taking flash photos to stop, but she kept clicking away. I asked the soldier to shoot those idiots, but he didn't understand English. The lights and activity on the wall eventually overwhelmed the hyena, who turned and disappeared into the bushes after spending only a minute and a half slurping butter from the mounds of porridge. So began the waiting.

According to the legend, the hyena was supposed to be joined by others and the porridge was to be eaten with only a little left; otherwise, the year to come would be pretty grim. But there on the ground lay three largely uneaten mounds of porridge, and there were no hyenas in sight. I scanned the surroundings with my camera and made out some glowing eyes about fifty meters away in the shrubs, but they remained where they were.[7] They seemed to be squabbling over something on the ground. After about a quarter of an hour, a dog arrived and began eating from the mounds of porridge, unaware of the sacred tradition he was desecrating. A well-aimed stone from one of the soldiers persuaded the dog to go away. After another quarter-hour, two soldiers came out of the compound and went across to where the hyenas could be heard squabbling in the dark. They found what the hyenas had been squabbling over: a cow's rear end. The soldiers dragged it back with them to a place beside the mounds of porridge. Shortly after that, two hyenas emerged and dragged the cow's rear end back into the shrubs, where they continued to squabble over the prize. More waiting. Inexplicably, a man emerged from the compound, went over to the shrubs beside the porridge, and urinated. I had no idea whether he believed that urine attracted hyenas or whether he had simply ventured out to relieve himself, but it made no difference. No hyenas came. After an hour of waiting, the murid decided that enough was enough and went down to examine the remains of the porridge. He bent over and looked closely at the three mounds before returning to the compound to organize the animal sacrifices. An ox and a sheep were led out of the compound, past the uneaten porridge, to a place north of the shrine, where they were slaughtered and butchered. Now, this did attract some hyenas, and a crowd of them gathered just beyond the lamplight. But they were fended off by dogs who positioned themselves between the slaughter and the hyenas. The heads, skins, and entrails were thrown out into the darkness, and the hyenas eventually displaced the dogs and dragged the bits back to the shrubs, where they squabbled and ate. After the noise died down, we all retired to our respective rooms for the celebrations.

Back in the men's room, the mood had changed. The drums were out and men were dancing to the driving beats, singing *zikris*, traditional Harari religious songs. The words of the songs were a mix of Harari, Oromo, and even some Amharic, generally praising a particular saint, in this case Aw Nugus. The copious amounts of khat that the men had been chewing gave them the stamina to keep the songs going all night. At about 1:00 A.M., there was a short break so that the murid could give a speech on the importance of Ashura and make predictions for the coming year. He said that based on what the hyena had eaten, a prosperous year would follow, and those who had come from Harar would be victorious. I suspected that the second part might have been added by Nureddin, who was translating, as he was preparing to run for office in the months to come.[8] The speech was followed by a supper of cooked meat and porridge with butter, served up in bowls identical to those used for the hyenas' porridge. It felt like indirectly sharing food with the hyenas. After eating, I went outside to see what had become of the porridge that had been set out. I went to the wall and heard the sounds of hyenas on the other side, but when I looked over the wall, they scattered into the shrubs. I shone my flashlight down on the stone slabs and saw that the porridge was almost finished. Maybe it would be a good year after all.

The singing and dancing continued until morning. By sunrise, I'd managed to fight off the urge to sleep, documented as much as I could of the ceremony, and was more than ready to go home. But I was beginning to worry that there would be no vehicle to take us back to Harar. The driver who had brought us to Aw Nugus had no inclination to stay for the all-night celebrations and had left during the evening. Nureddin was unconcerned about getting a ride home, though, and insisted that I go with him to see the healing ceremony that was taking place beside the women's room. While people sat and watched, three elders swilled a khat/water mixture and blew it over people who presented themselves wanting to be healed. The mixture and the gesticulations of the healers were intended to expel bad spirits. When three middle-aged women presented themselves for healing, one of them disrobed to the waist. Nureddin whispered to me, "This is a traditional Oromo practice which is contrary to Islam. It used to be commonplace, but now only a few women do so." As a professional anthropologist, I should have asked questions, taken notes, and delved further into this persistence of the old religion, but all I could think of was my need for sleep and the sounds of vehicles outside driving off. I urged Nureddin to make sure that we weren't left behind.

Before the healing ceremony was complete, we got up and went out the gate to find one remaining vehicle, a twin-cab pickup, surrounded by a score of people clamoring to get on board. Nureddin spoke at length to the owner and then hustled us into the backseat, while four more sat in the front and about ten in the truck bed. Like the last plane out of a war zone, we trundled down the hill with a trail of desperate people running after us, grasping at the back of the truck before giving up and resigning themselves to a ten-kilometer walk to Kombolcha. Yet the desperation was diluted somewhat by a gaggle of local children chasing after the car, shouting and laughing as they ran. The festival of Ashura must have been the highlight of the year for those kids. We rode back to Harar in tired silence, arriving at 8:00 A.M., and, after being dropped off in Jugol, returned to our respective homes to sleep.

Two days later, I met up with Mohammed Sayeed and told him about the footage I'd taken of the hyena eating porridge. "Oh ho, Mr. Kangaroo, this is a very amazing thing you did. You must have *waraba nasib*." Hyena's luck. I guessed I was lucky, but the video ended up causing me more trouble than it was worth. The cameraman from Harari TV was dissatisfied with his own footage and asked if he could have a copy of mine. I initially agreed but later realized that my ethics clearance required that I edit the video to remove footage of any people. When I told Nureddin of the cameraman's request, he phoned the cameraman and flatly refused on my behalf. This caused the cameraman to complain to the manager of Harari TV, after which Nureddin and the manager had an argument over the phone. Nureddin accused the manager of reneging on his promise to provide a car, while sending his own crew to Aw Nugus. He said they did not deserve a copy of the video after the way he'd behaved. The manager, who was a member of the ruling party, told his party colleagues about this, and within a week I was deemed an associate of Nureddin's opposition party.[9] The situation came to a head when I bumped into the director of the Tourism, Sport, and Culture Bureau. He embraced me and congratulated me on my success before fixing me with his stare and saying, "I want that video. If you don't give it to me, you will not leave Harar." Nureddin laughed this off as an idle threat, but I was pretty uncomfortable with my new status as an enemy of the government. I made a quick edit to remove any footage of the people present and left a DVD of the porridge-eating hyena with the director's secretary.[10]

In the aftermath of the Ashura festival and in light of the subsequent hype generated by Harari TV and the local newspaper, I was vexed by one question: if the hyena porridge feeding at Ashura was such an important element of Harar's cultural heritage, then why were so few Hararis at Aw Nugus on the night of the celebration? Looking back at the video, I estimated fewer than 120 people at the shrine. Of those, the majority were local Oromo who climbed the hill to participate in the celebration, half of them children. In all, only about thirty-five to forty people had traveled from Harar, and seven of those were my own party, who would otherwise not have gone to the shrine. Our driver had to stop twice to ask directions because he only vaguely knew the location of the shrine. Of a population of twelve thousand Hararis in the town and thousands more in rural areas and nearby Dire Dawa, this was a very paltry showing for such an important event.

Granted, few Hararis owned vehicles or had the means to travel independently the twenty-eight kilometers to the shrine. There was also a lot of celebrating in

Harar on that night, and many Hararis would have preferred to stick around for that.[11] Not to mention the khat factor. Once people in Harar settle down to an afternoon of burcha, it is pretty hard to motivate them to do anything, let alone journey to a shrine at the edge of the territory. Yet during the festival of Abdul Qadir Jailan, later in the year, convoys of buses ferry people to and from the shrine dedicated to that particular saint for a similar celebration, only without hyenas. So why not the same interest in Ashura at Aw Nugus? The bottom line is that the hyenas were the ones who determined the way in which the festival of Ashura was maintained and celebrated.

Feeding porridge to hyenas to ensure a prosperous new year is a precarious business and one over which the human participants can have little control. All the humans can do is to try to create the right circumstances for the hyenas to come and eat the porridge. The rest is up to the hyenas. Thus the murid put controls in place at Aw Nugus: the children who would otherwise have thrown rocks and chased away the hyenas were herded into a square and guarded by soldiers who had been brought specifically for the occasion; the majority of people attending were made to stay indoors after the porridge had been left out so as not to frighten away the hyenas. The murid was proactive in creating the right circumstances for the hyenas to come and feed, and this involved having as few humans as possible present to witness the feeding. It's a simple equation: the chance of hyenas coming to eat porridge is inversely proportional to the number of people present. So, from a Harari's perspective, what's the point of attending the celebration when you'll be made to stay inside, where you won't see a thing? No point, really, and considering the possible options—not being allowed to see the hyenas or hyenas not coming to feed—it's little wonder that the majority of people in Harar maintained the festival as a legend rather than an ongoing practice.

I suspect that the way in which the Ashura festival was maintained was cyclical in nature and varied across shrines. Conceivably, a shrine where hyenas came to eat porridge was going to increase in popularity. Harar is a small town, and word can travel fast. Harar also has TV, radio, print, and social media to amplify the gossip factor.[12] So, without controls, you could expect increasing numbers of people to go to any particular shrine that was successful in the hyena feeding. But once the numbers became too high, the hyenas would inevitably withdraw their participation, and after a few years of this, the number of people attending would decline to just a few.[13] At that point, the hyenas would return and start eating the porridge again. Hence the murids of the various shrines were in a position where they had to maintain the celebrations and yet try to ensure that the number of people didn't prevent the hyenas from coming to feed. When the hyenas abstained, it was the murids who had to deal with the failure.

But what kind of failure are we talking about here? When the hyena emerged from the shrubs and began to eat the porridge at Aw Nugus, the potential for the ritual to fail was massive. This was specifically due to the actions of two people: the journalist with the flashlight and the girl with the digital camera. The hyena was extremely nervous about the flashlight, and this was exacerbated by the intermittent flashes from the camera. The journalist and girl either failed to recognize or disregarded the mindset of the hyena—that she was scared and about to scarper—and continued with the flashlight and the camera until she ran off without finishing the porridge. This undermined what the murid was trying to do: make the place as unthreatening as possible to the hyenas so that they'd come and finish the porridge.[14] Unlike the journalist and the girl, the murid was attending to the subjectivities of the hyenas, and this was going to be key to the ongoing success of the Ashura celebration at Aw Nugus. The hyenas dictated the terms under which they would come and eat porridge, and it was up to the humans to understand those terms and accommodate the hyenas. The failure of the ritual in this case was a failure to attend to the subjectivities of the hyenas. If the hyenas declined to eat, then the implications for the coming year were ominous and obvious to all.

A few weeks after Ashura, I was sitting on the step at the Sofi feeding place as usual, watching tourists and hyenas. A vehicle arrived and three farenjis emerged, arraying themselves around Yusuf. A well-dressed Ethiopian accompanied them, but he had obviously seen the hyena feeding before. He ignored the hyenas and took a seat next to me at the shrine. He said to me in very clear English, "You must be Mr. Marcus." I was surprised that a complete stranger knew my name. The man introduced himself as Ahmed Zekaria, the very person who wrote the 1991 article on Ashura. He had come from Addis with some European visitors whom he was hosting. "I heard that you made some video of the hyena eating porridge at Ashura. Congratulations for that, it is very remarkable." I told him about the events and the kerfuffle over the video. We sat and chatted about each other's research and what I was learning about Harar's hyenas. After some time, the people he'd brought to the feeding finished taking photographs and indicated to Ahmed that they wanted to leave. Before he got up, I asked, "Did you really manage to film the hyenas eating porridge at Aw Aboker?" He smiled. "We went with the film crew and waited and waited. There were too many people there. The hyenas never came. In the end we brought some porridge here to this place and filmed Yusuf's hyenas eating it. The film is still in Germany waiting for someone to do the editing." Ahmed stood, bade me goodbye, and climbed aboard the minibus with the Europeans. As the vehicle withdrew and the feeding place became dark, the hyenas rushed in like a tide and scoured the ground for bits of food that had been overlooked.

# 6

## ON THE TAIL OF A HYENA

After Ashura, I really knuckled down and focused on my research. I constructed some questionnaires and, with the help of a translator from Haramaya University, conducted surveys in five villages west of Harar in Oromia State. Interestingly, the rural Oromo people held attitudes and beliefs about hyenas very similar to those of the Hararis.[1] This spoke of a flow of ideas across ethnic and stone divides.[2] This was even more the case in the Argobba village of Koromi, where they practically adored hyenas. They told me that if a hyena attacked a livestock animal, then they would have to leave out more food so he wouldn't be as hungry next time. In Harar I recorded the events of the festival of Aw Aslahddin. Some women elders camped out for a few days at Suqutatberi, making coffee and raising money. On the third day they slaughtered an ox, distributed the meat, and put out the entrails for the Sofi hyenas to eat. Strangely, the hyenas were afraid to come, so we carried the offering to the playing field at Abadir School and listened as entrails were devoured in darkness. I was also making a determined effort to make observations of hyenas in the Old Town. Night after night, I sat opposite one of the hyena holes near the feeding place and waited for a hyena to go into Jugol. Night after night, I found my chin dropping onto my chest as I struggled to stay awake during my fruitless vigils. The hyenas were simply not doing what I needed them to do; they weren't going into Jugol—or at least not through the entry holes that I was watching. I'd encountered a few hyenas when I was walking in the lanes and tried to follow them, but they made it very clear that they were uncomfortable with a human on their tails, so I'd had to back off. The only hyena whom I managed to follow and observe at any length was Bouki, but he'd made only a single brief foray through a hyena hole and into the Old Town. He investigated a couple of open compounds and then exited, leaving me with about ten seconds of data in my voice recorder. At the time, I didn't even recognize him as Bouki—it was only when I checked the spot patterns against my hyena photos that I identified him—so I ended up with a commentary on an unknown hyena entering and exiting Jugol, having done pretty much nothing. I went on like this for a couple of months before I had any success. In the end, my efforts were rewarded by the young hyena I mentioned earlier, the one who has a special place in my heart and a photo on my wall: Kamareya.

Some of the names that Yusuf gave the Sofi hyenas bespoke a considerable fondness for those particular individuals. Tukwondilli means "to touch him is a sin"; Koti means "mine"; and Chaltu means "refined." In the case of Kamareya, the name means "like the moon." I never asked Yusuf why he gave the young male hyena this particular name, but I guessed that it was because of Kamareya's roundish face and large eyes.[3] When I was beginning my observations at the Sofi feeding place, Kamareya was an adolescent, about the same age as Baby. In fact, he and Baby were close friends and used to hang out together at the feeding place while everyone waited for tourists to arrive and for the feeding to commence. While the two young hyenas were about the same age, I doubted that they were littermates because they looked so dissimilar. Whereas Baby was dark, with small eyes and a narrow face, Kamareya was all roundness and light-colored fur.

The differences between Baby and Kamareya were not limited to the superficial. While Baby was charismatic and fascinating to watch, Kamareya was quite inconspicuous. In fact, Kamareya verged on the dull. He was as audacious as Baby and stood as close to Yusuf and the tourists as she did. He even stood close to the seat at the shrine, waiting to steal morsels from the food bucket when it was unattended (or when I was supposed to be guarding it and pretended to look the other way).[4] But despite his audacity, Kamareya was nowhere near as demonstrative as Baby. He acted and reacted in a very low-key manner and never really gave away what was going on inside his head. He was like the kid at school who hangs around with the very popular or troublesome kid, more notable for his associations than in his own right. In short, Kamareya was a bit of a nobody.

Whereas I'd gone backward in my habituation attempts with Baby, I made progress with Kamareya. Initially, he took his cues from Baby. When he saw that she was unperturbed by my close proximity, he too became accustomed to lying down near me. Second, his fondness for the feed bucket led him to spend a lot of his time in front of the shrine, in close proximity to Yusuf and myself. If he was not standing in front of it, waiting for a chance to sneak in and steal some food, he was lying down close by, facing in the direction of the bucket. He became predictable in this, so I figured that it could work to my advantage. I began sitting on the pile of rocks beside Yusuf's fence in anticipation of Kamareya's coming over and lying down next to the rock pile facing the shrine. After Yusuf finished a feeding session and the tourists had gone, he returned to the shrine with his feed bucket. Kamareya followed and lay down in front of the bucket, thus very close to where I sat on the rock pile. In this way I was able to get closer and closer to Kamareya without making it explicitly obvious; it was Kamareya's choice whether he wanted to lie down as close to me as he did. And he ended up getting very close indeed. After all the tourists had gone and Yusuf retired to the house to finish his khat, I found myself sitting so close to Kamareya that I could hear his breathing. With each intake of breath, he made a little wheezing sound, audible only to the habituated.[5]

I was interested in where Kamareya was setting his boundaries with regard to me, where the limits to his tolerance lay. So I started seeking him out and approaching him. I found that if I was careful and pulled up short every time Kamareya looked like he was a little too disturbed by my approach, I could get closer and closer; I'm talking centimeters here. I reached the point where I could approach and sit within touching distance without Kamareya's doing much more than lifting a cautious head. Once seated, I exhaled demonstratively, as a hyena would, and Kamareya relaxed and returned his chin to his carpals, eventually even closing his eyes. This turned me into an unwitting sideshow one night when a friend of Abbas's brought some Ethiopian tourists around to the back of Yusuf's compound and pointed out the farenji and the hyena sitting together. Kamareya and I were both pretty disturbed at having a flashlight directed at us and a bunch of tourists taking photos.

Two nights after that incident, I found Kamareya lying down on the side of the road outside Yusuf's. It was late, so there was little chance that a truck or Bajaj would drive down the road and disturb him. He could relax. I walked up as though it was the most normal thing in the world and sat down beside him, eliciting no reaction other than a half lift of the head. A brief glance at me and he lowered his head again, while I shifted my position and leaned down on one elbow. At that he rolled onto his right side so that his body pulled a bit further away but his paws extended almost to my arm. I looked at those paws, with their waxy, bristly fur, and wondered at the texture. I had the compulsion to reach out and touch them. I began tapping on the road, a little beat. Tap, tap, tap, a dozen times, only a couple of centimeters from Kamareya's paws, while he watched my hand as if transfixed. Then one wayward tap and I touched his paw.

There was sensation and violent reaction all compressed into a fraction of a second. In an instant, I felt Kamareya's fur, the toughness of his hide, and was surprised at how hot it was. I never knew hyenas ran at such high temperatures. But my millisecond of musing was interrupted by Kamareya's explosive recoil at my touch. In an instant he jumped sideways and landed a meter away, wide-eyed and wary. This startled me, and I found that I'd recoiled about a meter in the other direction and landed with my heart racing. We stood there on the road, eyeing each other and catching our breath. I immediately regretted what I'd just done and thought I'd better give him some space, so I turned and went down the road toward Suqutatberi, leaving him in peace. Walking along the road beside the ancient stone wall, I kicked myself for upsetting Kamareya like that, and I wondered how much of his trust I'd undone. Would he follow the same path as Baby and actively avoid me from then on? It seemed that the biggest obstacle to my investigating the private worlds of hyenas in Harar was going to be me.

As it turned out, I was even inept at having misgivings, and my fears that Kamareya would redouble his boundaries were completely unfounded. Even though we'd agreed that physical contact was unacceptable, we became close foraging companions in the Old Town. Just a few nights after the paw-touching incident, I was sitting on my rock pile as usual, waiting for a hyena, when Kamareya came ambling along from the direction of the feeding place. He gave me a sideways look as he passed by and then turned right and went straight into the hyena hole. I was up like a shot and hurried after him, fumbling with my recorder, flashlight, and camera as I ducked my head under the lintel and hurried into Jugol on the tail of a hyena.[6]

I had to move fast. Inside the hyena hole there was a common with lanes leading off in three directions. If I didn't ascertain which lane Kamareya had taken, I'd have a limited chance of finding him. Staring hard into the darkness, I made out his silhouette heading toward Suqutatberi. I went after him as fast as I could, but without breaking into a run. If he heard running footsteps behind him, he'd surely run off. I saw him make a right turn up ahead, so I hurried on to the intersection. I rounded the bend and there he was, standing in front of some steps, sniffing at the water that was trickling down. He looked at me briefly, as if to make sure it was me, and then continued sniffing. From where I stood, I could smell food, and so could Kamareya, but the scent was so mixed up with the water that he couldn't locate the food. He moved on. This time I was able to keep up, and we climbed some steps that took us to a four-way intersection. Just ahead, a dog was barking maniacally through a gate that was slightly open. Despite the racket he was making, the dog wasn't game enough to come out of his yard and confront us. I shone my flashlight, which unnerved him, while Kamareya went behind me and down a drainage lane.[7] The lane was like a minefield for the obsessive compulsive.

There were bags of garbage spilling their contents and little brown mounds where people had gone to defecate. I pleaded with Kamareya not to keep going in that direction but he ignored my pleas and continued. Then a pair of glowing eyes up ahead and another hyena emerged from down the drainage lane. In the darkness, I couldn't make out who the other hyena was, although she looked familiar and she must have known me because she wasn't too concerned about my presence. She and Kamareya sniffed each other and went together down the drainage lane. Looking down the minefield, I couldn't bring myself to follow, so I ran back to the intersection, took a left past the barking dog, and ran to the metaled road hoping to circle around and meet Kamareya where the drainage lane exited.

Running down the metaled road, I realized that I was sweating. Keeping up with hyenas was turning out to be hard work. I planned to pass the first lane and take a second to the outlet of the drainage lane, but I pulled up short when Kamareya appeared from out of the first. He took a look at me and then returned up the lane. I followed and found him alone in the middle of the drainage lane. This time I resolved to follow him no matter where he went, so when he waded through the brown water and clambered up the meter-high drop-off in the middle of the lane, I clambered up after him. When I cleared the top, I found Kamareya waiting for me. He turned, and we both headed back up the lane to where he found a bone. Good, I thought, a chance to catch my breath. I stood looking about me while Kamareya crunched on the end of the bone, accompanied by the crazed barking of the nearby dog.

About halfway into the bone, Kamareya stopped munching and looked toward the top of the drainage lane. I followed his gaze and saw two hyenas approaching warily. I thought I recognized one as Diraatu, but it was dark and I could only see them from in front, so I couldn't identify them by spot patterns. They looked like they were going to take Kamareya's bone, but before they got any closer, Kamareya picked up the bone and edged past me, putting me between him and the other hyenas. How did he know that this would fend them off? It felt a bit awkward, standing between those two hyenas and Kamareya with his food, but then the two indicated that they were wary of me, so I held my ground. They gave up on the idea of taking Kamareya's bone and headed back up the lane in the direction of the main market. When Kamareya finished, he looked at me, obviously wanting to get past. I encouraged him and he edged past me once again, this time crossing the four-way intersection and turning right into yet another drainage lane!

I begged him not to go that way but again he ignored me and went on up, wading through the brown water running along its center. I had to hop from side to side on the narrow ledges that ran alongside the drain, super careful not to lose my footing and end up in the murky soup below me. We arrived at the top, whence Kamareya led me on a zigzagging path all around the Argobberi quarter. We stopped briefly while he sniffed the rear end of a cow who turned to watch him but was otherwise unperturbed, and then headed back uphill and into the Suqutatberi quarter. I eventually lost him when he was sniffing around a common. While I watched from the entrance to the common, two people marched in from a side lane and Kamareya bolted. I went after him, but the two people were ahead of me, blocking the lane, and it took me some time before I could remember the Harari for "excuse me" and get past. By then Kamareya was long gone, but I was contented. I'd recorded an hour's worth of hyena observations in the Old Town.

The following night, I couldn't help noticing that Kamareya was being persecuted by the other Sofi hyenas. I told myself that it was bias—that I was just noticing it more because of the events of the previous night—but it was quite overt. When he tried to feed, he was growled at and attacked, and then, between feeding sessions, one group of three hyenas went shoulder to shoulder with their tails up, driving him all around the feeding place. Later in the night, when I was sitting near him, Dibbey got up and marched straight toward him, forcing him to get up and move off. No wonder the locals believed that hyenas held grudges. Still, I wasn't about to give up following hyenas on the basis of a fanciful notion that a hyena could get into trouble for associating with a human. Suddenly, the door to the secret lives of hyenas in Harar had opened a crack and let through a little bit of light. I was intent on finding out more, and had to wait only two nights for another opportunity.

As usual, I was sitting on the rock pile, fighting the urge to sleep, when Kamareya once again came from the feeding place, turned right, and went through the hyena hole. I was immediately up and on his tail. Once inside, he turned right and went over to where two camels lay tethered to a post. Without hesitation, he sniffed at the rear end of one of the camels. The camel turned to see who this predator was and looked a little put out but made no protest. Kamareya finished with the first and sniffed the rear end of the other, who pretty much ignored him. Having satisfied himself that the camels were healthy, or whatever, he went on into the maze, eventually leading me to none other than the series of drainage lanes. I was resigned by now to spending most of my hyena-following time in Harar's drainage lanes, so without hesitation I clambered up onto the side ledges and hopped side to side along the lane as Kamareya plied the central drain. He had his ears folded back, following my progress as I followed him. We arrived at the top and turned left along a couple of lanes that led us in the direction of the main market in the center of the Old Town.

On our approach to the main market, Kamareya stopped short of a lane off to one side. He listened for a while and then turned and came back toward me, sniffing apparently aimlessly at the ground. I wondered what this strange behavior meant, so I edged past Kamareya, who stopped sniffing the ground and watched me as I took a look around the corner. About fifteen meters down the lane I made out the silhouettes of two hyenas crunching on some bones. I shone my flashlight down the lane to make an identification and illuminated two subadults. This is odd, I thought. I was certain that I knew all of the subadults of the Sofi clan—there were only about five coming to the feeding place at the time—yet these two were completely unfamiliar. Perhaps they never went to the feeding place out of fear of the people. But they were unafraid of me. It didn't add up. I turned back to see what Kamareya was doing and found to my surprise that he was looking at me intently, as if waiting for me to do something. Then the realization hit me: Aboker hyenas!

I should say a word here about the kinds of dynamics that can exist between hyenas of different clans. As I said earlier, hyenas live in social groups called clans, which range in size from five to as many as one hundred individuals. These clans maintain exclusive territories that they mark with feces and with scent from their anal and interdigital glands.[8] In a place like the Namib Desert, where prey is scarce, the hyena territories are so large and the clan membership numbers so low that it's impossible for the members to mark and defend boundaries comprehensively. Instead, they mark core areas within their ranges.[9] In places like Ngorongoro Crater, prey is so concentrated that several clans consisting of large numbers of hyenas live in very close proximity. Their territories are delineated by strictly defined, regularly marked boundaries, over which the clans frequently engage in what are known as "clan wars." In many cases, these wars are triggered by prey animals. If a wildebeest is pursued by some hyenas across a territorial boundary and killed within the territory of another clan, that other clan will intercept the intruders and chase them back to their own territory. Hans Kruuk and Jane Goodall both documented several such incidents in which rival clans engaged in boundary disputes over errant prey.[10] Most of the time, these involved a lot of posturing and chasing one another about, without any serious injuries. But there were instances when hyenas a little too slow to run away were caught and severely mauled, and in some cases killed, by rival clan members. In other words, hyenas can take their territorial boundaries very seriously, and the penalties for intruders can be quite severe. It was this knowledge that was foremost in my mind when I found myself standing beside Kamareya in a dark lane, with two Aboker hyenas around the corner and no idea whose territory we were in. The hairs on the back of my neck were just beginning to bristle when I looked beyond Kamareya to the other end of the lane and saw a third Aboker hyena approaching.

This new arrival was not exactly large. She was a subadult of about the same age as Kamareya and the two Aboker subadults around the corner. But she was big enough and unfamiliar enough to really put me on edge. She was very interested in Kamareya and certainly edgy herself. Her tail was up, her muscles tensed, and her mouth open; all of these are indicators of a very aroused and potentially aggressive hyena. I wasn't about to wait around and see if she was going to engage us peacefully; I charged toward her, stamping my feet. Thankfully, she took off up the lane and watched from a safe distance. To my surprise, Kamareya hadn't moved. I thought he'd have run at least a little way off when I went stomping past, but he was unfazed. He turned away from the new arrival and took another look down the lane where the Aboker hyenas were. When I shone my flashlight on the two, they picked up their bones and headed down to the bottom of the lane. At that, Kamareya went and inspected the ground where they'd been eating.

I followed and waited while Kamareya began crunching on a bone that the two had left behind. In comparison to the previous few minutes of tension, this was a marked contrast. It was positively boring. In general, it takes hyenas a long time to chew through bones, and Kamareya, as a subadult, was especially inept. He wouldn't develop his full bite strength for another three to four years; he even found it difficult to get a purchase on the bone.[11] Partly out of interest in what he was eating and partly just wanting something to do, I moved closer and directed my flashlight toward Kamareya's face and the bone between his teeth. It was at that point that he did the most memorable thing: he growled at me and made a little head thrust in my direction. I instinctively backed off, saying sorry, switching off my flashlight, and giving him some space. I stood there in the darkness of the lane contemplating the implications of what he'd just done.

When Barbara Smuts discussed the habituation of baboons, she described it as a process in which she, as researcher, underwent the greatest amount of change. The baboons, according to Smuts, remained themselves, "doing what they always did in the world they had always lived in."[12] Yet in the case of Kamareya's and my habituating to each other, we'd both undergone a transformation. While I was becoming attuned to the ways of hyenas, Kamareya was becoming attuned to me. Consider his reaction should another human have marched up and pointed a flashlight in his face. Actually, another human would not have gotten close enough to use a flashlight; within the Old Town, Kamareya would have bolted if another human came any closer than twenty meters. But he was adjusting his actions toward me as a unique individual with whom he was engaged in an increasingly complex relationship. He was waiting for me at the ends of lanes, letting me stand between himself and manic dogs, watching for my responses to foreign hyenas, inserting me between himself and hyenas who wanted his food, and standing fast while I went stamping toward a threatening Aboker individual. Moreover, he'd opened up a line of communication whereby he fended me off from a piece of bone and made clear his annoyance at having a flashlight poked in his face. I'd stretched his patience too far, and he told me to get lost. These were not the typical things hyenas did with people in Harar.

This is where I want to bring Martin Buber back to the table, because I have a bone to pick with him about the capacities of nonhumans, specifically hyenas, to engage in meaningful, transformative relations with others. According to Buber, nonhumans are unable to engage in the kinds of intersubjective I-You relations of which humans are capable. The key, Buber says, is human language. Buber asserts that without abstract language, animals remain at the threshold of You-saying, unable to distinguish fully between subject and object and therefore unable to achieve the kind of self-actualization that emerges from the I-You relation. I take exception to Buber's valorization of abstract language because it is in fact an inhibition to the immediacy of Buber's You-saying. Abstract language is just that—communication by abstractions; it requires internalization and interpretation before the self is able to respond to the other. When Buber set up language as the uncrossable divide, on one side of which is the capacity for You-saying, he put humans on the wrong side! When it comes to engaging with the other as a significant being, in all the immediacy of self-actualization, you need to reach beyond abstract language and engage with the whole other through your bodily self. Only with the entire body can we comprehend the other as truly other coming toward self. This is the kind of encounter that engenders the I-You relation, and I can testify that hyenas are so much better at it than we.

For all his lack of charisma, Kamareya was in fact quite adept at You-saying across species in both explicit and subtle ways. After seeing my response to that first little growl and head thrust, he readily adopted those gestures in the dialogue between us. When I stood between him and a bone, when I was too intrusive with my camera or flashlight, and even when he wanted to drive me away from my camera bag and procure its contents of bread rolls, Kamareya growled and thrust his head at me with exaggerated lunges. These are forms of communication in the imperative, yet beyond imperatives. A growl says more than "get away!" If it were to say "get away," then the immediacy would be lost. The growl cannot be translated into words because it surpasses speech, which must be dwelled upon. The growl calls to the You with immediacy, and the You must respond with immediacy. There is no place for abstract language except in a stupid, meaningless apology on my part, and the only meaningful aspect of that apology is my tone of voice.

Really, Kamareya must have seen me as something of a simpleton in terms of communication. So far as I could tell, hyenas signaled their fears, desires, and intentions with much more subtlety than Kamareya used with me. It's only in hindsight that I can take stock of the other ways in which he was engaging me in conversation—such as his stopping at the end of lanes and turning to look at me. He didn't do this to ascertain who was following him; he knew from the sound of my footsteps and had tracked me along the lane with his ears swiveled backward. The stopping and looking was an invitation to follow, and a quick read of my body set told him that I intended to follow. Once satisfied that I was following—having read intentions that I was unaware I was declaring—he turned and continued on.[13] He could read my embodied intentions when I was fending off dogs, giving him space to pass by, or charging at other hyenas. Yet poor Kamareya had to tolerate this witless human idiot who failed to read anything he was communicating other than a growl and lunge, which were far more explicit than should have been necessary in the circumstances. Indeed, we humans are exceptional in our command of abstract language. It is something that markedly distinguishes us from all other species; it underlines our communicational ineptitude and constantly inhibits us from engaging in meaningful, intersubjective relations with members of our own and other species.

Sitting on the hill behind the feeding place, I could see Yusuf silhouetted against the taxi headlights, handing strips of meat to the hyenas. It was an unusually busy night, with many tourists and a lot of hyenas, including Koti and her entourage, who crowded around the hyena man. One hyena emerged from the fray and ambled in my direction. I could see that it was an adolescent and guessed from the relaxed way in which the hyena was walking toward me that it was either Kamareya or Baby. It was Kamareya. He took a look at me and ambled past to a spot on the ground a couple of meters away, where he dropped to the ground and exhaled demonstratively. As if seized by an irresistible urge, he rolled over onto his back and, with his paws hanging loosely above him, danced a little twist as he rubbed his back into the loose dirt. A few twists more and he rolled onto one side, facing me, and looked me in the eye for a moment. Then another twist and he held his position with his paws in the air before rolling onto his other side, facing away from me. I shifted my position so as to lean on one elbow, and Kamareya lifted his head with his ears swiveled backward. I said something like, "Don't worry, I'm just getting comfortable," and he lowered his head once again and exhaled. I leaned further back, closed my eyes, and listened to the night.

# 7

# ENCOUNTERS WITH THE UNSEEN

Nearly five months into my time in Harar, I was in my groove. I became increasingly familiar with the lanes in Jugol and navigated my way through the Old Town like a seasoned local. I bought khat daily from the khat market at Assumberi, where I bantered with Samira, my favorite khat lady. I spent my afternoons in one of half a dozen houses where I had an open invitation to burcha. Often, my hosts invited friends who could tell me meaningful things about hyenas, and I ended the afternoon with a full notebook. In the evening, I bought lentil *sambusas* from Fatiya, the sambusa lady at Argobberi, and took them to the feeding place, where I shared them with Abbas.[1] By this time, shopkeepers and market sellers knew me and treated me like a local, charging me local prices rather than inflated farenji prices. Even the local kids knew me and no longer shouted out "Farenjo!" when I passed by. One or two occasionally called out "Hey Marcos!" while the remainder largely ignored me and focused on whatever game they were playing in the lane.[2]

I became very attached to the Sofi feeding place. It wasn't just the considerable amount of time I was spending there or the ready availability of hyenas, nor was it the comfortable step in front of Aw Ansar's shrine or the broad shelter of the fig tree that grew through the building. There was the warmth of Yusuf's family. Yusuf's wife, Naime, with her chubby cheeks bespeaking her Somali ancestry, was welcoming and motherly. She was also physically expressive, as is typical of Oromo women from that part of the world; she used the full gamut of gesticulations characteristic of her people. When the family crowded into the living room to watch Ethiopian soap operas, Naime was more entertaining than the TV drama. She shook her hand at disturbing outcomes or clapped her hands by bringing one up to meet the other coming down. She gasped at dramatic unfoldings and constantly exclaimed "Allah!" over some insult delivered by one of the characters. All of this with a wide grin and a sideways knowing glance at me, as if she was enjoying her own characterization of herself. Her eldest daughter, Ardalle, was equally vivacious but more conservative in her demonstrativeness, always catching herself and restraining her reactions. Still, when the urge took her she wrestled with Nasruti, the next-oldest, and delivered remarkably violent slaps to Nasruti's head while the two of them laughed uncontrollably. Nasruti, with her one milky eye, was a classic middle child. She added the occasional witticism here and there but never took center stage. That was reserved for the youngest, Ajebbo. Lithe like her father, Ajebbo commanded attention, and I mean that literally: if I was listening to Nasruti, Ajebbo pulled my chin so that I'd have to turn and face her. She was constantly on the move, snapping her fingers and slapping her hands as she talked, increasing her volume if anyone else had the temerity to try to interrupt her. She became reserved only when Yusuf entered the room. When her father was present, Ajebbo sat quietly, almost shaking with barely controlled exuberance. Ardalle used to taunt Ajebbo, saying she was *jinnam*, which Ajebbo earnestly denied. Jinnam translates variously as crazy, uncontrollable, mentally disturbed, or possessed by a spirit, although none of these meanings in Harar is mutually exclusive. Each state can be attributed to the kind of entity from which jinnam takes its root: jinn.

It's pretty hard to gauge from the Quran exactly what kind of beings jinn (sing.: jinni) are, although most people in the West can get some idea if they think of a genie. Same thing. There's a chapter of the Quran (Al-Jinn) that is essentially an account of the jinn accepting the Quran and promising to be good Muslims, promising not to associate anyone else with God. But this gives no indication of what jinn really are. Ar-Rahman (55:15) mentions that they were created from smokeless fire, indicating a spirit essence in these beings. And in the Hadiths, the accounts of the life of Mohammed, we find that they inhabit waste places and can manifest themselves as snakes.[3] The rest is left to interpretations, which vary considerably but among which there is a common theme: there are good and bad jinn; the good ones can assist you on a path to spiritual enlightenment; the bad can deceive you and, given the opportunity, take possession of your being.[4] Thus someone in Harar who is jinnam is very likely to be under the influence of a bad jinni and as such should have reason to fear hyenas. This is because hyenas can see and hear jinn and, given the opportunity, will catch and eat these otherwise unseen beings.

"Jinn kabri zo iydenta waraba karsi!" The tomb of the jinn is in the hyena's belly! I was back at Abdulmuheimen's travel agency, sitting on one of the three-legged stools.[5] Abdulmuheimen was leaning across the table, with a cigarette held aside as if forgotten, while he looked me squarely in the eye and, stuttering, stressed the importance of what he was saying. "Hyenas are the vaccine of the territory of Harar! Whenever there is smallpox or swine flu or whatever it is, you just vaccinate that area. That way, smallpox will be contained. So even jinns are contained in the territory of Harar because of the hyenas. Whenever jinn comes, the hyena he is saying oowhoop, and there it goes." Abdulmuheimen pointed to his mouth to indicate the fate of the jinni. "He is the scavenger of the city, the Sufi scavenger of the city. So whenever he will come to the city and come in the waraba nudul, he will clean up the visibles and the invisibles." What are they like? Jinn? "They say that today, we cannot see them. Tomorrow, they cannot see us, but we will see them. On this world, we cannot see jinn, except very few people, okay? That's what they say. Man and jinn live together. We live together. There were three Sufis who were assigned from dawn to about this time and this time to about dusk. Three people until the time of the Italian invasion. They meet, they chased, they shepherd the jinns out of the city. In the daytime these three people, and in the night, the hyena will come and eat them or they will escape. If the jinns is under the earth, then the hyena comes and yells oowhoop and it will come into his mouth."

So there was meaning in that melancholy sound, and that meaning was reinforced by the act of making it. When a hyena whoops, he puts his nose to the ground and his whole body seems to wind up as the oowhoop emanates and resonates. Serengeti researchers Marion East and Heribert Hofer suggested that hyenas do this specifically to increase the resonance of the sound so that it might travel farther (hyenas can apparently hear one another from five kilometers away).[6] But why not conclude that a whooping hyena is sucking an unseen spirit from the ground? He certainly seems to be doing something meaningful and satisfying. One researcher friend of mine thinks that hyenas, once they start a whooping bout, are unable to stop—kind of like a bout of hiccoughs or an orgasm. He deduced this from seeing a hyena who, having started to whoop, was being chased by another hyena yet was unable or unwilling to stop whooping. There's been quite a bit of research done on hyena oowhoops; the sound is so conspicuous and captivating. Apparently, higher-ranking females whoop longest and males whoop most. Cubs whoop to get support from their mothers, while males whoop to advertise their presence to potential mates and other males.[7] In Harar, the oowhoop is just one piece of evidence of hyenas' capacity to eat jinn.

Down the road from the travel agency, I was at Hakim Café, visiting Tigist and showing her some close-up photos of the Sofi hyenas. Where she came from in West Shewa, hyenas were creatures to be feared, so she found it amazing that someone could approach hyenas so closely and not be eaten. One of the wait staff, a boy named Fahmih, insisted on sitting between us and practicing his English on me. Another person doing so would have been pretty annoying, but Fahmih was inherently likeable. He was thin, like most boys in Harar, with long hair in oily ringlets that made him look like a young Ethiopian Lionel Richie. He lived in Jugol but was actually a Muslim Amhara—a descendent of Amhara settlers since colonization. He also spoke excellent English despite his limited education, suggesting a natural talent for languages. I questioned him about jinn and hyenas. "You know the dead people place?" The cemetery? "Yes, that one. When a man is dead and they bury him in that place, a jinni comes up from his grave and goes into Jugol. If the hyena sees him, he makes a big noise, oowhoop like that, and makes the jinni afraid. He runs away and goes into the ground, into a hole. The hyena, he sees the jinni and goes up to the hole to make the noise again. Oowhoop! This makes the jinni come from the hole and he runs again, but the hyena, he is too fast. He can catch the jinni and eat him. After that, the jinni is *burka* . . . dead. Then, you know, the hyena goes away to a big hole but in the mountain." A cave? "Yes, cave. He goes into this place and makes vomit of the jinni onto the floor. You can go into the hole at this time because the jinni is dead, and inside there is all kinds of things to make you rich. There is gold and money and a special stone. If you have this stone, you can get sheep, goats, cows, even camels. And there is a special bowl. You can put a coin in this bowl and after there is many coins just like that." Fahmih was animated by this time, imagining the kinds of riches that could take him away from his dreary, impoverished existence into a world of cars, travel, electrical appliances, and unbounded leisure—the world from which I apparently came, and which ironically had no hyenas to regurgitate a jinni's riches.

At the eastern end of Hakim Café, just past the ragged beggars arrayed on the steps between two roads, stood the regional government offices and parliament. I was there at the invitation of a Harari man named Mohammed whose wedding I'd attended in Suqutatberi and who insisted that I tell the state president about my research project. We were admitted without having to wait and greeted warmly by the president, who beamed when I addressed him in Harari. He apparently held none of the petty antipathy toward me that some of his ministers held after Ashura; he listened with rapt interest as I told him about Ashura, the Sofi hyenas, and the progress I was making. As we talked, the discussion turned to the subject of jinn. With Mohammed translating, the president expanded on what I knew. "Jinn is everywhere. You can't see but they're all over. Good and bad, there are all kinds of jinn, just like people. The good jinn leave people alone but the bad jinn can . . . eat people." I could tell that Mohammed was looking for the right word for the way that jinn took possession of people. I suggested "possess" and he agreed. "When a bad jinni is over you like that, then you are afraid of hyenas." I asked if it might be the case that a hyena attacking a person was in fact a case of a hyena attacking a jinni that was "over" a person, and they both agreed. Then the disclaimer: "But a hyena won't do this in Harar." The president continued, "If a hyena sees a jinni, he will eat him up. After this, you know, he will vomit. There is a place outside of Assumberi where hyena goes to vomit jinn. You can go there and see the vomit. Inside is the fingernails and hair from the jinni and sometimes even the fingers. With these things you know the hyena has eaten a jinni."

There it was again. These people weren't just making this stuff up during burcha; they were paying attention to the hyenas and the things they did out there in the town. Unlike dogs, hyenas do not regurgitate food for their young, but they do gag up slimy balls of undigested hair and slivers of bone and hooves. These are the bits that hyenas' strong digestive juices fail to dissolve or else aren't worth dissolving. Other hyenas are also strangely attracted to these gag-balls, and an instant after a hyena vomits another is quickly on the scene, rolling ecstatically on the little ball of undigested bits.[8] They frequently do this around their dens, where cubs emerge and join in the self-anointment of slimy hyena vomit.

Back in Jugol, at the Arthur Rimbaud museum, I was talking to curator and friend Abdul-Nassir when we were interrupted by the night watchman, Khalifa. I liked Khalifa a lot. He was about sixty-five years old, with mottled teeth and eyes from chewing khat but with a keen exuberance. He was one of the few people who immediately grasped the extent of my command of Harari and adjusted his words and gesticulations to suit. He spoke not a word of English, but between his Harari, dumbed down for my benefit, and his hand gestures, which verged on mime, he could communicate whatever he wanted to say without need of a translator. He came to tell me about an incident the previous night at a house near the Italian mosque. Apparently, a group of women were preparing sausages for a wedding when a band of ten hyenas barged through the half-open gate and raided the kitchen.[9] Pandemonium ensued as the women ran screaming into the house (the kitchen in a Harari compound is usually a separate outbuilding) and bolted the door, while the hyenas helped themselves to the spoils. Khalifa knew something was up when he saw, in the lane outside the Rimbaud Museum, ten hyenas marching along as though they owned the town. Abdul-Nassir interjected at this point that there must have been jinn in the compound that attracted the hyenas. Khalifa asked him what he'd just said, and when Abdul-Nassir told him, Khalifa agreed. He explained to me that the Rimbaud Museum was in fact brim brimming with jinn. He said he protected himself during the day with Quran readings, and at night he hung up pieces of hyena skin in his watch house, which straddled the lane outside the museum. Apparently, the smell of the skin was sufficient to keep the jinn away. Encouraged by Abdul-Nassir, Khalifa produced from his coat pocket a square piece of skin with the distinctive bristly fur of a hyena. On inspecting it, I was immediately struck by the smell, which lingered on my hands for the rest of the afternoon until I got home and washed. Jinn or no jinn, the smell from that skin would be enough to keep anyone away.

I'd been introduced by Nureddin to one of Harar's former mayors, Zeyadan Bakri, a big man with a beard and warm eyes, which made him at once intimidating and strangely cuddly. Zeyadan was very keen on my hyena research; his was one of the houses where I had an open invitation to burcha. He was home most afternoons, chewing khat with friends and colleagues, so I often visited to talk about bureaucracy, colonization, cultural heritage, and, of course, hyenas. Zeyadan was erudite and worldly; it was under his stewardship that Harar won its UNESCO World Heritage listing and City of Peace Prize. One afternoon, the subject of jinn was broached, and Zeyadan related to me an account of his youth. "This was over twenty-five years ago. I was with my family when we heard a lot of noise and commotion from the house next door, so we went to see what was happening. The people next door said their girl was affected by a jinni. She'd been wanting to go out at strange hours, like midnight, and having trouble paying attention. They invited a sheikh, who was in the room with the girl and asking her questions. Why are you here? What do you want here? The girl was nervous; she wanted to run from the place but she was restrained. It took all three of her brothers to restrain her. She was yelling and giving strange answers to the questions, like 'None of your business!' and 'Get lost!' The sheikh made readings of the Quran for hours and after some time the girl fainted. After that, she slept for two to three days and then she was fine. She went back to school and was completely normal. At the time, I didn't understand what was happening—I was just a boy—but later I understood. She was infected by a jinni."

I had a chance meeting with Abdu Semed Idris, a Harari cultural historian, who featured in the Dermesheh video. We got onto the subject of jinn and he told me about a sheikh he knew called Haji Idri. When a couple visited the sheikh with their mentally disturbed daughter, the sheikh made Quran readings to effect a cure. During the healing, the girl began to speak in a man's voice and tried to make a deal with Haji. The sheikh refused the voice's deals, to the point where it agreed to leave if given ash. He fed the girl some ash, the girl screamed, and then she fell asleep. When she woke, she had no recollection of what had just transpired. She slept for a couple of days and then, over the following months, regained her appetite and the weight she'd lost. Abdu Semed confided, "I was an atheist at the time, reading Marx and Engels, but after the incident with the girl, I completely changed my beliefs."

I felt like I was walking a line between two kinds of the same entity. On the one hand, there were the jinn, which were tied up in the worlds of hyenas. These were harmful, mischievous spirits who lurked in waste places and entered the town at night. These jinn were subject to hyena predation. While normally unseen, they were suddenly made visible in the vomit of hyenas or when they appeared to people as humanlike. They were feared for their duplicity but not for their ability to possess. On the other hand, there were the possessive jinn. These were spirits whose first appearance came when they entered into the soul of a person—usually a young girl. These jinn took control of their subjects and left only after expulsion, normally after a sheikh's Quran readings; hyenas had little or nothing to do with them. Then one night at the hyena-feeding place, the two kinds were dramatically brought together.

I arrived at the Sofi feeding place a little late one evening and took a seat next to a young Amhara wannabe tour guide named Addisu. His English was good but he didn't have the front, the quick-witted cunning, or the immunity to rejection required for his trade. I often found him sitting at the shrine, watching the hyenas and hoping that some unaccompanied tourists might happen along and hire him. This was a vain hope; by the time tourists arrived at the hyena feeding, they were proficient at refusing offers from tour guides. On my other side sat a well-dressed Muslim woman. I guessed that she was from out of town, a member of the emerging Ethiopian middle class. Beside her sat a boy in fashionable clothes and sneakers, with his head turned away and crying to himself. I took no further notice of the woman or boy and began my hyena observations. Here, Funyamurey displacing Jaagi; there, a young hyena on the periphery, afraid to come and feed. But my focus was broken when Abbas's voice called out to me from alongside. He'd come from inside the house, followed by Elias, the farmhand, and was on his way past the shrine. Hey, where are you going this time of night? I asked in Harari. Abbas was grinning from ear to ear and replied, "Off to the farm to get some hyena poo." That stumped me, but before I could pursue it further, Abbas and Elias had vanished into the darkness.

By this time, the woman beside me was standing and had the boy by the arm. He was still crying and obviously didn't want to get up from the step, but Mum was a lot stronger. She literally dragged him over to where Yusuf was feeding the hyenas and stood behind the boy, holding him with both hands while he tried to get away. Poor kid, I thought. If he's afraid of the hyenas, why doesn't she let him stay back at the shrine and just feed them by herself? Yusuf glanced sideways at the boy and reached into the food bucket for a strip of fleshy skin. He took a stick and planted it in the boy's hand, while the mother held the boy's arm from behind. Yusuf draped a bit of skin over the stick, Funyamurey snatched it, and the boy squirmed and cried out. At that, the boy was allowed to drop the stick, and his mother led him back to the step. The boy calmed down a bit and sat quietly. What a way to enjoy a holiday.

When Yusuf finished feeding, he returned to the shrine and sat on the other side of Addisu. I asked through Addisu what Abbas was doing getting hyena feces from the farm, and the answer came back that he was going to make some medicine from it. That should be interesting, I thought, so I sat with Addisu and Yusuf in front of the shrine, waiting for Abbas to return. But when Abbas appeared with Elias, holding two handfuls of chalky white hyena feces, Yusuf remained sitting at the shrine. Abbas and Elias went into the house, this time followed by the Muslim woman and her boy. Not wanting to push the issue, I sat quietly and waited for Yusuf to get up and go make this medicine he'd spoken of. We waited fifteen minutes before Yusuf finally got up and went toward the house. I quickly got up and followed him to the door, where we encountered a man coming out. The man greeted me in French, then English, and went on his way, followed by the woman and her boy, who was by this time quite contented. Behind them came a smiling old man holding a Ziploc bag full of the hyena feces that Abbas had collected. They climbed inside their car and drove off into the night, leaving me in dire need of an explanation. I called Addisu and brought him into the house to interpret Yusuf's account of what had just transpired.

"The boy had a jinni," he said, as though this should have been obvious. "The family is Somali, living in Djibouti, and they tried there to find some help for the boy. The doctor said he couldn't do anything, so they took him to a sheikh. This sheikh said they had to take the boy to Harar and make him feed the hyenas. After that, they had to bring home some waraba *guff* and he would make some medicine." But I thought Yusuf was going to make the medicine. "No, it's the sheikh. Yusuf says this is not the first time people came here for this thing. A time before, a Somali family came with somebody who had a jinni. They came in the afternoon and made Quran readings. He says it's mostly the Quran that gets rid of the jinni."

So that was it. The two disparate strands of possessive jinn and crocutaphobic jinn had come together right before my eyes, and I hadn't even been aware of it. It was all so logical, too. Jinn are afraid of hyenas, so why not take the boy to a place where hyenas congregate and scare the jinni out of him? You didn't need to be a sheikh to figure that one out. And then there was the hyena feces. The stuff is of hyena and transports easily once dry. And it's easy to make into a powder, so why not use it as medicine for jinn possession? Yusuf said that even though the jinni had left the boy, it would try to return over the next several months. The powdered feces would be smoked in a clay pot and the boy made to inhale it until the jinni finally went away for good. This is how hyenas can cross boundaries. Not only can they see the unseen and hear the unheard; their bodies produce powder that turns into smoke and enters the spirit world to banish the jinn. Hyenas are powerful, fearful animals in so many ways, and if you're concerned about jinn, quite handy to have around.[10]

I'd witnessed an exorcism of sorts, but the jinn weren't finished with me. My friend Adil, the one who quit his job in Addis because of the hyenas, intercepted me outside Jugol Hospital one afternoon and took me aside. He said he wanted to introduce me to his friend Mohammed so that we could talk. I knew who he was talking about but wondered why he wanted me to talk to this guy. "You're interested in jinn?" Yes, I am. "Mohammed has one." One what? "A jinni." Okay, I thought, maybe I needed to talk to this guy.

It took a couple of weeks before I was eventually able to meet with Mohammed. He canceled twice before finally appearing at my door with Adil, who was to translate. Mohammed was thin, with a round head on the end of a long neck. A little stubble grew from his chin, and he had small round eyes between protruding ears. He looked a lot like a character from a Miyazaki film. I invited them to sit on the mats I had arrayed on the floor (furniture is rare in Harari houses). After passing them some khat and reading through my ethics declaration, I sipped some water from a bottle and pressed the red button on my handheld recorder. I was surprised at how quickly Mohammed broached the subject.

"The voices first came to me on the day of a funeral. We were all sitting around the mourning house, chewing khat, when someone switched on the radio. The station was playing Quran readings, but after a while it was too repetitive, so someone switched the radio off. The thing was, I could still hear the Quran readings. The rest of the night at the mourning house, I heard them, and they continued on my way home and when I went into my house. Then a voice began describing what I was doing. The voice said things like, 'He is sitting down. He is moving his right hand.' Things like that. I went to the toilet and that was when the jinni entered me. She went into my body and to one of my kidneys and said, 'I'll be here until the day you die. My love is for you.' I asked who it was and she answered, 'I love you.' I asked where she was from and she said, 'I saw you at Gidir Magala.' I asked her to leave but she refused.

"After that time, many jinn came to me. This was a really difficult time for me. Whenever one was about to come to me, I felt this wave coming over my body, and then the jinni would come. They asked me for sex or to go near to water, but

I refused them all.[11] They gave me information about what other people had in their hearts, what they were thinking but not saying. They told me when someone was going to knock on the door before the knock came. Then they started to get jealous. When I saw a pretty girl in the street, they asked, 'Why are you looking at her?' I tried to deny it; I told them I wasn't looking. Some of these jinn came from other people who I used to see in the street. They came to me at night and had sex with the jinni who was over me. I told her—the jinni who was over me—to get lost, but she just laughed. I said, 'If you don't stop, I'll fuck you up.' She laughed at me, so I found the woman who she came from and told her to stop disturbing me. The woman laughed and walked away, but that jinni left after that.

"There are Muslim jinn and Christian jinn and others. The Muslim and Christian jinn fight over my soul, but when bad jinn come, the Muslim and Christian join together and drive the bad ones away. The bad ones try to take health from my kidney, but the good ones fight them and restore my health. After that they start fighting again. If the Christian jinn make medicine for me, the Muslim jinn keep quiet and let them cure me. The reason they fight for my soul is that it can cross over to other places; it can travel over the ocean and visit faraway countries. It can cross religious borders too. Normally, the Muslim jinn keep the Christian jinn away, though, and it is they who protect me from *sheytan*.

"Whenever I go out, the jinn use animals to talk to me. I see a snake or a lizard and it asks me, 'Where are you going?' They change their behaviors. I see a cat and he says, 'If you don't do what I say, I'll kill you.' I'm sure I'll have to answer a lot of questions after talking to you today. They'll be angry with me. Right now there are snakes in the room crossing from here to there and going under the door. They're saying I have a lot of trouble for this. The only animal who the jinn don't talk through is waraba. He feels for you. When I'm trying to sleep at night, the waraba comes and talks to me with the oowhoop. He makes this sound and it's like thunder crashing down for the jinn; they all run away and hide. Waraba is like an angel from heaven coming to me and sending away the jinn."

I didn't know what to say. I felt bad for getting Mohammed into trouble with his jinn. Adil was stunned. He knew that Mohammed had a problem with jinn, but he had no idea how pervasive it was, how debilitating, how consuming. As if hearing our thoughts, Mohammed said that any other person experiencing what he experienced would go mad. He could deal with it, though, because he knew how to keep them away. "You have to be honest and you have to pray." I asked if he wanted me to delete the recording. He thought for a while and said yes. But he let me hang on to the notes I'd written during the interview.[12] He said his good-byes and departed, leaving both Adil and me to ponder what it must be like to experience the world in such a way. It was late afternoon and several hours before the hyenas began patrolling the lanes. I tried to imagine cats and lizards speaking to me, internal voices taunting me, and the tug and pull of spirits competing over my soul. Imagine having to deal with that every day. And then I imagined a hyena passing by my house, and that haunting, melancholy "oowhoop" from just outside in the lane, reaching into the room—into me—and, for a brief time, making it all go away.

# 8

# REFLECTIONS FROM A HYENA PLAYGROUND

My fifth month in Harar saw the arrival of a new hyena at the Sofi feeding place. He was about seven months old when he first arrived, sandy-colored, with his spots well established. What he found must have been, for a young hyena, a most surreal scene: a human, illuminated by vehicles and surrounded by other humans, sitting on the ground handing out food to hyenas. For many young hyenas, this scene was all too intimidating, so instead of getting involved, they lay down on the hill cloaked in secure darkness. They only ventured over to the feeding area when the people had gone away and the lights were out. This new arrival, though, succumbed to his fascination and joined his seniors at the feeding place, where he stood on the periphery, wide-eyed, trying to make sense of the spectacle. Like a wrestling referee, he dodged about the place, watching carefully every bit of food given over and every hyena altercation, yet without getting directly involved. Hyenas are incurable rubbernecks; they watch everything that goes on so as to keep track of the social organization of the clan, they join other hyenas in fights with dogs and intruders, and they capitalize on any opportunity to get food. After a while, the young hyena deciphered the machinations of hyena feeding and approached to feed.

The night that this young hyena first accepted food, I was sitting on the step of the shrine in conversation with an Ethiopian tourist. Abbas, who was feeding, called out to me in Harari from the middle of the feeding area, "Hey, Marcos! The new hyena is feeding!" I looked across and saw the youngster standing—almost hiding—amid three adults and tentatively approaching Abbas to take the food that was being reserved exclusively for him. The adults tried to get at the food, but Abbas kept pulling it away until they held off and allowed the youngster to feed. I called out to Abbas in Harari, What is the name? And Abbas replied, "It's Willi." At the time, my command of the Harari language was pretty limited. I was unaware that *willi* was in fact not a name at all but the Harari word for "early." Abbas was telling me that it was too early to give the young hyena a name, that he'd think of one later. In my ignorance, I continued to call this hyena Willi, and Abbas, who thought my misunderstanding hilarious, began using the name himself. Consequently, when Yusuf was feeding and asked Abbas the name of the new hyena, Abbas told him "Willi," and the name stuck. Not only was Willi the only hyena with a Harari name, but he proved to be a very unusual hyena in many other ways.

I think that Willi made a misguided assumption about me based on what he'd encountered at the feeding place. It was there that the majority of his encounters with humans occurred, and in these encounters it was not only the hyena man but every other human who offered food by turns. And this was especially the case with the fair-skinned humans. As for me, I was human, fair-skinned, and often at the feeding place. It was only logical to expect that I would at some point hand over some food. Willi began to follow me around. On the night of his naming he followed me over to the hill when I went to sit with the hyenas. He stood and looked at me for some considerable time before he was distracted by some hyenas squabbling. The following night, I encountered him as I was walking up toward Argobberi. He followed me halfway up the road before giving up on me and following someone else walking down toward the feeding place. Regardless of the number of times I failed to provide food, he continued to show an interest. It seemed that it would take some time before he associated me with disappointment, as Baby initially did. But then Willi took our association to a different place.

A month after Willi's first feeding, I was sitting on the pile of rocks across from the hyena hole when he wandered onto the road. A couple of people walked past on their way to Suqutatberi, and he followed them for several meters before they realized that they were being followed by a hyena. They found this a bit too disturbing, so they stamped their feet and chased Willi off, whereupon he returned to the stretch of road near where I was sitting. I spoke to him in a soft voice and he approached me, bobbing his head. This is a habit of young hyenas faced with strangers, lions, or even blades of grass waving in the wind. I held completely still while he sniffed my flashlight and then my hand. I'd never felt a hyena's nose before. He turned his attention to my knee and sniffed that for a bit. Then he turned his head sideways to my knee and opened his mouth. I thought a bite was imminent, so I quickly swung my knee away and reprimanded him. That was enough to send him scuttling off a few meters, whereupon he sat and washed himself.

Watching Willi and thinking about how he'd attempted to bite my knee, I was struck by the same question I'd had about Tukwondilli: what does he think I am? Did he seriously think that we humans were walking piles of meat, like gyros in a kebab shop, from which he could just bite off chunks of flesh? Did he miss the part of hyena feeding where the meat came from a bucket, not from the bodies of the hyena man and the tourists? Or did he actually consider us intentional subjects, but ones who willingly took strips of our own flesh and proffered them to hyenas? This in turn raised all sorts of questions for me. Hyenas are predators. Certainly, it would ease the conscience to conceive of a wildebeest as a very fleet yet mindless object in the landscape, but wouldn't a theory of mind be useful in catching a creature who was determined to avoid being eaten? But then, hyenas are pursuit predators, so maybe they can get by without ascribing any intentionality to their prey. They just have to chase it until it slows down enough for them to disembowel the thing.[1] Still, Willi was young, and as with a human child it might have been just a matter of time before he understood the intentionality of others and how that included the universal desire not to be eaten. Regardless, Willi's interest in me was undiminished, and he persisted in his attempts to establish some sort of relationship, even if that precluded eating bits of me.

On the fourth night after that biting attempt, I went around to the rock pile beside the road and found Willi and Baby lying down nearby. When I sat down, Willi got up and came directly over. He again sniffed my knee, but this time without trying to bite. Then he sniffed other parts of me, even my nose. Before long, Baby got up and joined Willi, and I found myself the center of attention for the two hyenas. Having had a good sniff, Willi turned his attention to Baby. He opened his mouth and pressed it against Baby's neck in a similar fashion as he'd done with my knee. I wondered if I'd got it wrong and he hadn't intended to eat my knee at all. Then Baby opened her mouth and pressed it sideways against Willi's back. This they did right in front of me. The sniffing and mouthing ended when two unfamiliar males arrived at the feeding place and Willi and Baby went over to intercept them with their tails erect.[2] But this night was just a precursor to what was to follow. Willi was about to begin habituating hyenas on my behalf.

At the feeding place, Willi was becoming ever more insistent on some sort of engagement. On one occasion, after the feeding was over, I found myself sitting in the dark among the hyenas, riding out a rain shower. Willi, Hadha Kamar, who ranked about fourth in the Sofi hierarchy, and I were sheltering in the lee of Yusuf's fence, which leaned over sufficiently to fend off most of the rain.[3] I'd thought hyenas were impervious to rain, the way they normally lay out in the open and seemed unperturbed by the water, but apparently some preferred to remain dry. We all lay still, staring at nothing in particular and letting the water wash over us. It was at times like these that hyenas showed me how to contemplate the moment.

When the rain diminished to a sprinkle, Willi got up, stretched his limbs, and wandered over to the garbage dumpster, where Baby was sniffing around. The two young hyenas sniffed each other and then began pushing each other with open mouths. Before long, Baby was biting at Willi's neck while Willi twisted and turned, making it difficult for her to get a grip. Then, as if someone had fired a starter pistol, Willi was off and running across the hill with Baby in pursuit. He led his pursuer across the face of the hill and down onto the dirt road before turning and racing up toward the fence, Baby hot on his heels. It was exquisite to see hyenas running at full speed across rough terrain like that, and I doubted that many animals could keep up—or get away. Once back on the hill, Willi stopped and turned to face his pursuer, who set about biting his neck. He twisted his head to break free and the chase continued. Another circuit of the hill, and then Willi did something very strange: he ran directly toward me. I sat still, not knowing what to do, as he came to within a couple of meters before veering off and running down toward the road. Another bite from Baby down on the road and Willi once again ran straight toward me. This time, he stopped in his tracks about a meter in front of me, and Baby all but crashed into his rear end. The two of them stood with their tails up and without hesitation began sniffing me. Baby tried to circle around behind while Willi sniffed my knee and opened his mouth. I felt his teeth against my knee and it was all I could do to stop him biting me. Meanwhile, Baby was sniffing at my backside, making me decidedly uncomfortable but at the same time enthralled with my predicament.

With teeth coming at me from both sides, I was compelled to stand up. Normally, a standing human would have been at least a little intimidating to a couple of hyenas, but Willi and Baby were unfazed. Rather than being intimidated, they seemed to take it as an invitation to pursue their quarry and tried to bite my legs. I thought I should enlist another human to my side, so with two hyenas snapping at my heels I hurried around to Yusuf's gate. Arriving in the compound, I called to Abbas to come and see what was happening, but by that time Willi and Baby had broken away from me and gone charging off once again onto the hill. Abbas called out to me from the house, asking what was happening out there. I replied, Nothing! and ran back toward the hill, where Baby and Willi were engaged in biting each other. I ran toward Willi and reached down to try to grab him but he dodged out of my way. I noticed that he wasn't showing the least sign of fear in response to this but was deftly avoiding me, stopping just a few meters off and waiting for me to charge at him again. I turned and charged at Baby and she did the same, while Willi circled around behind me. At this, I squatted and pivoted around as Willi and Baby tried simultaneously to get behind me and take a bite. I thrust out my hands at the two of them and tried to grab a handful of fur, but both were too quick for me; they dodged about and stayed out of my grasp.

While I was fending off the two hyenas' biting attempts, Kamareya came bounding up from the feeding area with his mouth hanging open. He darted toward Willi and snapped at the youngster's tail before veering off and running down to the road. We all three turned and chased Kamareya across the face of the hill to the dirt road below. Willi chased him into the forest across the road from the hill, while Baby and I stood at the edge watching the two. They emerged from the forest and raced down the road toward the farmland with Baby and me in hot pursuit. The three young hyenas engaged in some roughhousing at the bottom of the road before Willi broke free from the group and ran toward me to do some more biting. As spontaneously as it began, the play fizzled out when Baby's attention turned toward some interesting scent or other at the bottom of the hill. Willi went and lay down, while Kamareya wandered off toward the farmland. I headed for home. Walking up the hill toward Argobberi, I experienced an indescribable high, the kind of feeling that cannot be conveyed in words, that comes from engaging in play with three hyenas on a grassy hill under the light of a full moon.[4]

The very next night, Willi spun the wheel. I arrived at the feeding place just as the rain was easing. Yusuf was at the farm, guarding against khat thieves, Abbas was in the house with the rest of the family, and the hyenas were lying about the place with their chins on their carpals. The only activity was the darting to and fro of the night birds under the canopy of the fig tree.[5] The seat of the shrine was wet, so I went over to the far side of the feeding area and sat on my heels among the hyenas, adjusting my eyes to the semidarkness. A young hyena got up immediately and ambled over to me. It was Willi. Without hesitation he began sniffing my night-vision scope, my umbrella, and then my knee. I touched his back briefly and spoke reassuringly. He continued to sniff me, so I patted his back again. I kept expecting a reaction like Kamareya's when I touched his paw, but none came. So I put my hand on his head and rubbed the fur between his ears. Unfazed by my touch, Willi happily sniffed my arm and showed no inclination to bite. I called to Abbas in the house, who came out and pulled up at the gate when he saw me with Willi. He shook his head, and said, "Amazing, Marcos." I stood back as Abbas approached with his hand outstretched to pat Willi's head, but Willi recoiled. Abbas backed off and I was able to pat Willi once again. Abbas called the rest of the family outside. Naime and the girls appeared at the gate, where they all stood watching the strange hyena making friends with the equally strange farenji.

The following night, I arrived at the deserted feeding place and Willi came up to me straightaway. When he tried to bite my arm, I could see that he was going to be gentle, so I let him have at it. He felt his way with his teeth along my jacket sleeve, and then, when he managed to get sleeve and no arm between his teeth, he bit and pulled. At the same time, I grabbed a handful of the skin on his back and pulled at that. We wrestled like this for some time, each trying to get a purchase on the other. Later, he did the same thing with Kamareya and Baby, as they all three lay on the ground biting at one another's necks and legs and pulling on tufts of fur. It looked like pretty rough play, and on occasion one of them uttered a croak to indicate that the biting had gone too far, but none of them lost their enthusiasm for the play. At times, Baby broke away from the wrestling match and bounded toward me with her mouth open, veering away at the last moment as if testing my nerve—or her own. I refrained from chasing, as most of the play involved wrestling on the ground, which I would only disrupt if I tried to get involved. But the three hyenas were very aware of my presence on the hill, and every now and then one turned his or her head to look at me before another grabbed the skin of their neck and pulled them back into the fray.

The weeks after those three nights of play were not as carefree. Willi was still interested in biting, and he even gave me a nasty bruise when he bit too hard on my arm, but Baby and Kamareya were more interested in feeding or relaxing than running about the hill. On one night, a farmer brought his ox, who had died of disease, and with Yusuf's approval dumped the carcass in front of the shrine. The result was a spectacular hyena banquet wherein more than three hundred kilograms of meat and bone were transformed into a dark patch on the ground in less than an hour. After that feast, the hyenas were noticeably fatter and completely uninterested in attending the feeding. On the first night after the ox-fest, Willi was the only hyena at the feeding place, and he was only interested in biting me. On the second, no hyenas at all showed up until after 8:00 P.M., when Dibbey arrived. Yet she wasn't interested in feeding. It was as though the hyenas had a collective hangover and had all sworn never to eat again. Still, Willi persisted in attending and in biting my arms and equipment.

Nearly three weeks passed before we played again. The feeding was over for the night and the hyenas were arrayed about the feeding place and the hill. I found Willi behind Yusuf's compound, lying down with a bone through which he was struggling to chew. Just as I sat down next to him, Baby appeared from the road, bounding toward me with her tail up and her mouth hanging open, looking, to be frank, like an idiot. She circled around behind me, so I turned and lunged, reaching out to grab her, but she dodged my grasp and darted off up the hill. At about twenty meters distant, she wheeled around and came charging in again, barreling into Willi and biting his neck. Willi hadn't been particularly interested thus far, but Baby's bite made him stand and take a bite of her. They dug their paws in and pulled at each other until Baby gave in and allowed Willi to pull her to the ground right in front of where I was sitting on my heels. I knew that Baby was aware of my presence and wondered why she was unconcerned with me. I reached out and grabbed a handful of the fur on her back and pulled, at which she lifted her head a little but otherwise showed no concern. They both stood and wrestled their way toward the canal, leaving me where I was, but the instant Willi loosened his grip on Baby, she turned and charged toward me again. She veered away at the last moment and went running over to Kamareya, who was practically asleep, biting Kamareya's neck and tumbling over him. I got up and accompanied Willi to where Baby and Kamareya had each other in a vise grip. But as I approached, they managed to move away from me, edging farther and farther up the hill. Throughout, Willi shuttled between the biting hyenas and me, bridging our two species with his neck bites and sleeve tugs. As on the first night, the play fizzled out without any specific indication that it had ended. Kamareya was diverted by some interesting scent and Willi wandered over to the fence, where he lay down.

Often, the sound of hyenas can be heard echoing through the lanes of Harar on a Saturday afternoon. This is not due to any hyenas' getting their timetables mixed up and arriving before dark; the sound is actually that of Harari children mimicking hyenas while they play a game called Waraba. The mode of the game is a curiously commonplace predator-prey performance known to English speakers as hide-and-seek or hide-and-go-seek. In the Harari version, the child who is the waraba closes her eyes while the other children run off to find a hiding place. Once ready, the hiding children call out to the waraba to come looking. She searches the nooks and crannies of the lanes and open compounds while making the haunting "oowhoop" cry of a hyena. Meanwhile, the children who are hiding hold their breath and listen to the approaching beast, deciding whether to stay put or make a run for it before the waraba gets too close. If a hiding child is discovered, then the roles are exchanged; the one who was the waraba must go and hide while the other becomes the waraba and the game begins anew. Sound familiar?

For ecologist/philosopher Paul Shepard, games like Waraba are multifaceted in both meaning and purpose. On the one hand, they speak of the cyclic nature of the universe. The predator not only catches the prey but assimilates him. Hence prey becomes waraba and so too waraba becomes prey by assimilation. On the other hand, adopting the perspective of predator or prey is an exercise in the establishment of self-identity. By becoming the waraba, a child finds within herself the part of her person that is like the waraba. The self is projected onto the external world of animals, and correspondences bounce back to become incorporated into the identity of the child-person.[6]

But there is more to play than awakening the player to life as a game or the establishment of identity through mimicry. Play by definition is not "for real." In play, the rules of the game of life are temporarily suspended; the boundaries between beings are blurred and the status quo set aside. This is why we have so many ways of signaling that "this is play." It's because we need to signal our intention and arrive at an agreement with our playmates that whatever stood before is temporarily set aside and whatever happens during the game is part of the game.[7] Hence a child playing at Waraba is unconstrained by convention. Not only is her identity opened up to incorporate another kind of being, but her childhood restrictions are in some part removed. She can oowhoop like a hyena and stalk another human with the intention to capture and eat. She acts outside the narrow rules of her life as a human child and subscribes to the greater rules of life and the universe. In this way, play is both immediately and ultimately transformative. Once unconstrained by convention, the child can explore relations with others hitherto out of bounds. It is through play that we can chase and be chased, eat and be eaten, without reserve. This in turn leads to transformation beyond play, where boundaries between playmates are a little more diminished and possibilities greater.

Prior to Willi's intervention, Baby had been standoffish with me. She largely avoided me and showed little inclination to engage. However, Willi led her into a realm where she was open to some kind of transformation. As we played—as we took on the roles of chaser and chased—the boundaries between us became permeable. Suddenly, Baby was fully attending to me and my actions, as well as to my position within our circle of players. At first, I was prey, the horned, hooved beast in the middle, around which she circled, trying to find an opening. But then Kamareya came charging up from the hill and took a bite of Willi, and as if he'd flicked a switch, Kamareya became the prey, while it was left to the rest of us to pursue. We chased Kamareya to the road beside the forest, whereupon Willi followed up the pursuit while Baby and I stood and watched. I remember looking at Baby standing right beside me and being struck by how strange and yet how perfectly normal it was to be standing next to a hyena, staring into the forest. Suddenly, there was no species boundary between Baby and me; we were playmates and co-hunters of whatever it was that Kamareya was being, standing together in pursuit of our quarry. This was a different relationship from any we'd had previously.

On the night of play that followed the ox-fest, the relationship was again transformed. Baby instigated the play, and she was open to all kinds of possibilities. When I reached out and grabbed a handful of her fur, she was entirely aware of who was grasping her. Yet she let it happen. While previously (and subsequently) any kind of physical contact was entirely unacceptable, within the realm of play the previous boundaries became blurred, the rules were suspended, and everything became open to possibility. But the transformations that come with play sometimes resonate long afterward. About a month after that first night of play, I heard dogs barking down on the road and got up to follow the three hyenas, including Baby, who went to investigate. One of the dogs ran at Baby, but I intervened and chased him off with a bluff. Baby appeared at my side and we chased the dog down into the forest, where she walked just ahead of me with her tail up and her ears directed straight ahead. The feeling of comradeship was just as strong as when we chased Kamareya into the same forest. But this time we were not playing around.

The nights of play continued but were never really as exuberant as the first. Mostly, the play consisted of lying down and biting, in which case it was only Willi who involved me and only Willi with whom I could wrestle. Baby and Kamareya just couldn't drop their boundaries to the same degree.

As for Willi, he didn't need the circumstances associated with play to diminish the species distance between himself and me. He reached the stage where he could simply walk up to me and begin biting. Sometimes I gave a little hyenalike croak to indicate that he'd bitten too hard. He'd relax his jaw a little and allow me to pull my arm away, leaving him with jacket sleeve. If Willi wasn't interested in playing, he came over and lay down beside me, sometimes with a leg draped companionably across one of mine.[8] Otherwise, he might just give me a nip on his way past where I was sitting. The other hyenas paid attention to Willi's relaxed relationship with me and in turn altered their reactions to my presence. Kamareya and Baby both became more trusting and sometimes sought me out for company. Even the older hyenas put aside their mistrust and often lay down very close to where I sat with Willi on the hill behind the feeding place. Sitting among the hyenas, I began to feel a sense of belonging with them. Their familiarity and their acceptance of me brought a feeling of quiet companionship among all the spotted bodies of the Sofi clan scattered about the newly green hillside. I felt like one of them. Hence I felt a niggling complicity when I received a phone call from Nureddin telling me that four children in Kombolcha had just been attacked by hyenas.

**FIG. 1**    View of Harar's Old Town from the Orthodox cemetery.

**FIG. 2**    Harar Gate, the entrance to Harar erected after colonization. Andenya Menged and Jugol lie beyond the gate.

**FIG. 3**  A hyena hole in the Assumberi quarter.

**FIG. 4**  A hyena hole high enough to admit humans. Note the bits of timber halfway down, which suggest that this was once a smaller hole exclusively for hyenas.

**FIG. 5**  An Oromo farmer with a *muncha*.

**FIG. 6**   The shrine of Aw Ansar at the Sofi feeding place.

**FIG. 7**   The shrine of Emir Nur in Assumberi.

**FIG. 8**    Worshippers at the Orthodox church in Feres Magala going through the women's entrance.

**FIG. 9**    Yusuf feeding Koti, the highest-ranked Sofi hyena.

**FIG. 10**   Abdulmuheimen outside his travel agency in the New Town.

**FIG. 11** Aboker hyenas gathering outside Sisay's butcher shop. The cross painted on the wall indicates that the shop sells meat for Christians.

**FIG. 12** Bajaj and another Aboker hyena in the New Town.

**FIG. 13**  Aboker hyena backgrounded by rows of khat.

**FIG. 14** An old postcard of Tana feeding hyenas outside the wall. Photo courtesy of Nebil Stationery.

**FIG. 15**    Saloman feeding Flopsy of the Aboker clan.

**FIG. 16**   The billboard that Mulugayta had erected in Feres Magala.

**FIG. 17**  Diraatu, a high-ranking Sofi female.

FIG. 18    Chaltu, another of the high-ranking Sofi females.

FIG. 19    Bouki in the early morning sun.

FIG. 20　Yusuf's son, Abbas, feeding Funyamurey.

**FIG. 21** The skittish Sofi youngster Ibsa.

**FIG. 22** The intimidating Dibbey with her head lowered.

**FIG. 23**  Tukwondilli beside Yusuf's fence.

**FIG. 24** Baby at the garbage dump.

**FIG. 25** Nureddin wearing a *gey coloyta*, a traditional Harari cap worn at weddings and special occasions.

**FIG. 26** Kamareya on fallow ground east of the garbage dump.

FIG. 27   Baby plying the central channel of a drainage lane.

**FIG. 28** Naime, Yusuf's wife, sitting in the cooking hut making *ukhat*.

**FIG. 29**   Yusuf's eldest daughter, Ardalle, among the maize at the family farm.

FIG. 30   Yusuf's second-eldest daughter, Nasruti.

**FIG. 31**　Yusuf's youngest daughter, Ajebbo, picking the fruit of the prickly pear cactus (*tini*).

FIG. 32   Hadha Kamar in the waiting room.

**FIG. 33**  An early photo of Willi.

FIG. 34   Willi and the author on the hill behind the Sofi feeding place.

**FIG. 35**  A hyena killed by a truck on the outskirts of Harar. The ear has been cut off to be used for sexual potency and the skin from the eyebrow has been taken to create an amulet for protection against *buda*.

**FIG. 36**  Koti, Dibbey, and Willi marching up Suqutatberi Road toward the main market. Note Willi tucked in behind Dibbey's rear.

**FIG. 37**  Jalla watching a farmer from among the maize.

**FIG. 38**  Tigist being investigated by a curious Willi.

**FIG. 39** Flopsy and Big Red of the Aboker clan on their way home in the morning.

**FIG. 40** Young Willi at the garbage dump.

**FIG. 41**  Muka Dibbey, probable offspring of Dibbey.

**FIG. 42**  Fintamurey on the ground at the Sofi feeding place.

**FIG. 43** Doing "obs" in one of the research vehicles of the Michigan State University hyena research project.

**FIG. 44**   A collared hyena of the Talek clan in the Maasai Mara.

**FIG. 45**  Sofi hyenas foraging for food among the ubiquitous plastic bags at Harar's garbage dump.

**FIG. 46** Willi approaching some *waraba derbisi* (hyenas' wild fruit). This tart yellow fruit closely resembles another, called *derbisi*, which is consumed by humans for gastrointestinal problems.

**FIG. 47**   Kamareya standing on the irrigation channel at a point where it bridges a stream.

 **FIG. 48**  Going home.

# 9

# DEATH, DEATH, AND RHETORIC

After only a few months in Harar, I'd become the go-to man for any hyena-related news coming out of the Hararge region. When a hyena was hit by a khat truck racing through Arattenya in the early morning, I was inspecting her body by 11:00 A.M. When a hyena in Dire Dawa ran into a police station, seeking refuge from stone-throwing children, I fielded three or four phone calls from people who told me all about it. The same with the wedding-sausage raid and any other hyena-related incidents. Hence when Oromia Radio reported that there had been a series of hyena attacks in Kombolcha, I was taking phone calls all morning. Nureddin was the first to tell me. "Marcus, have you heard the news of these hyena attacks? I just heard on Oromia Radio, there is some conflict or other between the people and hyenas in Kombolcha." I remembered Kombolcha from our drive to the Aw Nugus shrine for the Ashura ceremony. The town lies at about three-quarters the distance to the shrine; we stopped there to pick up khat. "According to the news, some hyenas killed an ox belonging to a farmer. This farmer took some poison and put it on the body and the hyenas came back to eat it. Ten hyenas died from this. The very next day, some other hyenas came and exacted revenge. They attacked the village, killing some children."

I knew Oromia Radio well. The building was just across the road from the place where Tigist worked, where I used to go all too often to meet Mohammed Sayeed for coffee. I called Mohammed, who worked for Harari Radio next door, and asked him for his help. He took me next door and we went from office to office until we found the news editor.[1] The editor reiterated what Nureddin had told me—that there had been a series of attacks in a suburb of Kombolcha and that children had been killed and injured. They gave me the contact info for the Oromia police media liaison and suggested that I get the full story from him. I began to feel strangely like a journalist rather than an anthropologist and wondered how relevant these attacks were to my work in Harar. In any case, I called the media liaison and asked about the attacks. It turned out that he understood little English, so I enlisted my translator friend Mahir and went to visit him at the police commission. There we found a casually dressed man who looked far more like a media person than a policeman. We sat in the shade of a jacaranda tree and he related some of the details of the hyena attacks.

The conflict began when a group of hyenas broke into a farmer's hut where he was keeping his goats. The hyenas killed and ate several goats. The irate farmer applied some poison to one of the carcasses and left it out the following night. The hyenas returned to the farm and ate the carcass, after which ten of them died. This was apparently the catalyst for the attacks on the children. The hyenas returned to the village in the morning, seeking revenge. They attacked four children, of whom two were killed, one being eaten down to the waist. On seeing my reaction to this news, the liaison smiled and said, "Relax. It's just a kid." Another two children were bitten before the hyenas returned to their dens. The following morning they returned and attacked five more children, biting their heads and necks. In all of these cases, villagers threw things and chased away the hyenas before any of the children were killed. The morning after that, the hyenas came to the village again, but this time the people were ready and waiting. They chased the hyenas away and managed to kill one with a muncha. The hyenas returned once more and attacked two more children. The villagers again fought back; some of them were injured by the hyenas and taken to the hospital.

The following day, many parents in the village kept their children at home or else escorted them to and from school. The district administrator was at a loss as to how to respond. He called a meeting with the town councilors, police, elders, and religious leaders to try to arrive at a solution. The elders argued that the traditional response in such circumstances was to send a person who knew hyena language to negotiate a truce. This did not surprise me given what I'd found during interviews with rural dwellers. Like Hararis, the rural Oromo held that certain people could understand hyena language. They also believed that hyenas had a highly developed revenge ethic. Accordingly, any farmer who killed a hyena could expect to suffer the wrath of the hyena's clan mates, who would return and attack him and his family, or at least trash his farm. Hence finding someone who understood hyena language and negotiating a truce rather than retaliating fit neatly with local conceptions.

The police in Kombolcha, though, were reluctant to try negotiating with the hyenas. They argued that the situation urgently required a solution; there was no time for negotiations, let alone finding someone who could talk to hyenas. In the end, the district administrator decided to destroy the dens that were nearest the village. In this way, the hyenas would have no place from which to launch daytime attacks. Meanwhile, the farmer responsible for poisoning the hyenas had gone into hiding. Not only was he wanted by police for the illegal killings, but he also had to answer to the townspeople for inciting the hyenas to attack their children.

After telling me as much as he knew about the hyena attacks, the media liaison gave me a photocopy of the press release and said that everything he'd told me was written in the release. I thanked him and he said something in Amharic that Mahir was reticent about passing on. When I pursued the matter, Mahir smiled awkwardly and said, "He asked if he could have fifty birr for khat."

The liaison had told us that one of the children had been taken to Harar for treatment, so I went with Mahir to Hiwot Fana Hospital, guessing that of the four hospitals in Harar, it was the one to which an out-of-towner was most likely to be taken. The hospital itself was terrifying. In the dark corridors, floors halfheartedly mopped lay beneath black grime where they met the walls. Gravely ill people lay disconsolate on wooden benches waiting for hospital beds, while cries and groans emanated from darkened rooms foggy with disease. It was a horrifying and very likely place to die.

I'd forgotten to bring my magic letter from the director of the Tourism, Sport, and Culture Bureau, so we slammed immediately into administrative stonewalling. Our inquiries were met with responses of "Who wants to know?" and "Try coming back next week when the other administrator is here." Mahir eventually found a health worker who could tell us about the child. Apparently, she had head injuries and an infection that was being treated. He said the infection smelled different from normal infections. I remembered the piece of hyena skin Khalifa had shown me. When the health worker asked if I wanted to meet the little girl, I thought for a while and then declined. The poor kid was probably traumatized from having her face ripped open by a hyena. The last thing she needed was a farenji to come nosing around wanting her to recount the whole terrifying episode. We left the hospital and its patients to their fate.[2]

I wanted to go to Kombolcha to find out more about the attacks, but I procrastinated for a month before doing so. I'd learned to dread encounters with bureaucrats, and a visit to Kombolcha would be like going to a different country. The town was in Oromia state, so my magic letter would be useless. I figured I'd need a whole bunch of permissions, from zone level downward, and it took me weeks to summon the resolve to seek them. Once I'd determined to do it, though, it proved surprisingly easy. I called Ahmed Aliye, the East Hararge Zone administrator to whom Nureddin had originally referred me when I did my surveys north of Harar. Ahmed arranged for me to meet the following day with Wordi Hashim, a local administrator for Kombolcha. Wordi gave me his phone number and said, "Call me tomorrow when you arrive in Kombolcha."

The next morning I was riding along a bumpy road between green hills of khat on my way to Kombolcha. I was in my usual rental car, a Suzuki owned by the manager of a fabric shop named Abdulmalik who included his services as driver in the deal. In Ethiopia, it's extremely unwise for a farenji to drive a vehicle; if there's any kind of accident, farenjis are held responsible by default.[3] Presumably, this is because of our ignorance of local driving customs and our perceived capacity to pay for any damage. In fact, it was likely that I'd be held responsible for any accidents simply because I had hired the car and was sitting in the passenger seat. Riding in the back of the Suzuki was my translator and friend Arif. A Somali by extraction, Arif was a polyglot par excellence, counting English, Harari, and Oromo among the six languages he spoke fluently. He also looked uncannily like a young Haile Selassie, only his beard was a little narrower.

Kombolcha is a type specimen for Ethiopian B-towns. There are no historic landmarks; there is no town center; there is in fact no evidence of any town planning, just buildings lining either side of a dirt road and other buildings hiding behind them. Some are made of concrete blocks with tin roofs, most are wattle-and-daub with tin roofs, and everything is built on the cheap. This is due more to necessity than miserliness. We went to the *wereda* administration, a concrete building built and furnished in the fashion of an old school, and met with the administrator, who took us to see the wereda leader.[4] He greeted us, wound up the meeting he was in, and took us to his office, furnished with a modest steel-frame desk and little else. We sat and heard his account of the attacks and the underlying reasons behind them. "There are boundaries which we must not cross. People and hyenas must live in peace together, but the farmer who killed the hyenas crossed the boundary. So the hyenas, they were taking revenge for their friends that were poisoned." I asked what the solution was, and whether they had closed the school. "No, the option of closing the school was raised, but we didn't close. The solution was to destroy the hiding places of the hyenas. After we did that, the attacks ceased and things settled down." We thanked the wereda leader. He introduced us to a short, smiling Amhara man named Simmey, who worked for the administration. Simmey was instructed to show us around the town and mediate between us and the townspeople. I was beginning to formulate a theory that distance from the capital correlated with helpfulness in government officials.

Simmey took us to the northern end of the town to a place called Maalka Raafuu, where he said the original poisonings occurred. As we got out of the car,

Arif indicated some lantana and said, "This is *afkolo*. This is the hyenas' hiding places that the wereda leader was talking about."[5] So it wasn't the dens that were destroyed; it was the weeds in which the hyenas sought cover. In the wider world of knee-jerk responses to attacks by wild animals, this struck me as something of a sensible, rational solution. Simmey indicated some huts beneath a stand of eucalypts. "That is the farm where the hyenas were killed. Would you like to go and see?" I felt disinclined to go snooping around a farm. My presence had already drawn a crowd, and I didn't want the farmer to think I was investigating hyena killings with a view to prosecution. One of the local villagers piped up and offered to take us to the farmer, who was in hiding. This sparked a heated conversation among the locals, after which the villager withdrew his offer, explaining that he could get a lot of trouble from that farmer if he did so. I wanted to explain that I was a humble anthropologist and not some international observer come to prosecute the wildlife killings, but I knew that the people would form their own opinions. Whatever I said to allay their suspicions would be interpreted as lies intended to allay their suspicions. They'd only become more suspicious, so I simply said nothing.

Another heated discussion among the locals followed, and then a teenager stepped forward to talk to Simmey while gesturing to me to follow. We began walking along a path between huts while Simmey explained, "He's taking us to a hyena." When I asked him to expand on that, he said, "There was a hyena here this morning. It was at the school, hiding in the hedge. Some people saw it and they chased it all the way to here, where they killed it. He is showing us the dead hyena." As he finished his sentence, we arrived at an open area at the back of a cluster of huts where there was a conspicuous pit. We walked to the edge and saw, in the bottom of the pit, the body of an adult female hyena. A large gash from a muncha ran vertically down her bloodied face, and one of her ears had been cut off. That severed ear would be used for "sexual enhancement." Some pieces of skin had also been cut away. "Spiritual protection." A bloodied wooden club had been thrown into the pit with the body as well as some empty cigarette packets and some stones. Patches of her fur were burned from attempts to set fire to the carcass.

Standing at the edge of the pit, I wondered about this hyena's last moments. I imagined her flushed from the hedgerow beside the school and running through the village, surrounded by hostile humans, every one of them intent on her death. At some point her escape must have been cut off, and she was set upon by villagers with whatever weapons they had at hand. It was a terrifying last look at the world for a hyena. I immediately felt uncomfortable standing at the edge of the pit among a crowd of gawking, giggling humans. It felt ghoulish and disrespectful; I felt as responsible as every other person encircling the pit. I jumped in to be with the hyena.

I skidded down sideways so as not to land on top of her and stopped at the bottom. She was short and stocky like Dibbey. I reached down and lifted her leg to see her underside. Her leg felt strong, heavy, lifeless. Under her thigh, I could see two protruding nipples, like wallet-sized photographs informing me of her cubs. Like any other hyena, she had a life story. I asked Arif to translate, Why did they kill her? The answer came back, "Because the hyena was near to the school. It was late in the morning." But didn't the wereda leader say that the hyenas would seek revenge? "Yes, they are afraid of this. But the hyena was near the school and they thought it might attack one of the children." I could tell that Arif was also grappling with the inconsistency between the villagers' beliefs and their actions. It occurred to me that a crowd of fearful people intent on killing a hyena, for whatever reason, might not always pause to think of consequences. I felt uncomfortable under the collective gaze of the score of people who had gathered around us, so I was glad when Simmey suggested that we go to where one of the children had been attacked. We returned to the car and drove to another part of town.

We arrived in a newer suburb of Kombolcha where the wattle-and-daub houses had shiny tin roofs and clean red mud covering their walls. Driving to a cul-de-sac, we got out and were immediately surrounded by locals and children in abundance. Simmey spoke to one of the locals, and a little girl of about four years was brought forward. "This is the girl who is bitten by hyena," he said. His misuse of the present tense was quite apt, as she still bore the mark of the attack. Her face was swollen on the right side and she was missing some teeth.[6] The poor girl was confused and uncertain of why she was the center of attention. I asked where her mother was, and the reply came back that she was away at work. The girl was left in the care of neighbors, which was essentially no care at all, as she was simply running around with the other kids. Our circus was centered too much on the little girl, so I tried a diversion. Can anyone tell us about the attack? At this, a woman stepped forward and began talking in the uninhibited, expressive fashion of the Oromo, with one hand on her hip and the other waving about vigorously. Arif translated as she exclaimed and gesticulated. "She says the hyena came in the morning from over there." The woman was pointing to a stand of eucalypts. "The children were shouting, Hyena! Hyena! She looked and the hyena had the girl in its mouth. She says she ran after the hyena and grabbed him. That was when he dropped the girl and ran away." Without lowering his tone, Arif said, "I think she's lying about chasing the hyena. She wants to make something big of herself." It didn't matter. I wasn't interested in who saved whom or what caused what. In fact, I didn't really know what I was interested in. I felt I needed to be there and to try to understand something, but I was being confounded by dead hyenas, crowds of people, and neighbors big-noting themselves in front of a farenji. I did, however, know that I should get away and leave the little girl to herself. I asked Arif to tell her that it was good luck to be bitten by a hyena. I didn't expect the girl would understand, but maybe her neighbors would believe the visiting farenji and treat her just a little more respectfully, not in spite of but because of her disfigurement. The circus continued to grow, so we got back in the vehicle and drove off. The crowd parted down the middle as we drove away.

We went back to the wereda administration, where we met with an administrator who had all the details of the attacks. Muqtar was the man's name. He had a big nose, a serious demeanor, and was obviously busy keeping the town from falling apart. But when he sat down behind his desk, he gave the impression that at that moment our needs were the most important in the world. Muqtar took a piece of paper and drew a rough map of the town, on which he marked the locations of all the events connected with the hyena attacks. "Up here, the farm and the place where the hyenas were poisoned." How many hyenas were poisoned? "Eight altogether, but four of them were killed by the villagers while they were sick from poisoning. Over here is the cemetery. The hyenas live here and come to the town. The road is the boundary between two groups, called Dinni [Religion] and Kamar [Moon]. They fight on the road. The Kamar hyenas are the ones attacking people. Down here is the place where the first attack." Muqtar pointed to a place behind the administration building, about a kilometer from where the hyenas were killed. "This was a girl. They found her eaten down to here." He held his hand at the level of his chest. "The next one was in the morning here." Very close to the first. "The hyena took a child, but when he put her down another child chased him off. The third here." Within a hundred meters of the first two. "Two boys were visiting a relative and the hyena took the smallest. Some people saw this and chased the hyena. The boy was only injured. The fourth was this place." That was the little girl we'd met in the new suburb. "Again people chased the hyena. After that one, we had a meeting and decided to destroy all the afkolo so the hyenas had no place to hide. People were told to watch their children, and the attacks finished. That's it."

So the police press release was exaggerated. I asked Muqtar whether someone had suggested sending a person who knew hyena language to arrange a truce. "Yes, absolutely. But in the end, we decided it would be quicker to destroy the afkolo. If that didn't work, then we could consider other things." It was consistent with beliefs held elsewhere in the region that the administrators were taking care not to incite further hyena attacks. Rather than treat hyenas as objects in the landscape that could be managed with the right policy, the administrators were treating them as intentional subjects. Relations between the hyenas and the people had to be carefully negotiated, whether through direct communication or indirect action. It struck me that direct action against the hyenas was something the villagers might take in the spontaneity of the morning, but it was a last resort for the administration.

I wanted to know what happened to the farmer who poisoned the hyenas, so we went to the police station at the southern end of town. It was yet another single-level concrete building, this one in the shade of some poincianas. We were greeted by an inspector, a typically slender Oromo man with a prominent nose and cheekbones. Like most police officers in Ethiopia, his uniform hung loosely on him and his boots seemed too large—like a boy wearing his older brother's hand-me-downs a few years too early. The inspector was genial and helpful; he explained that the farmer had been caught and prosecuted. In addition to killing wildlife without a permit, he was prosecuted for creating a disturbance and for not disposing of poisoned animals. The penalty imposed was a five-thousand-birr fine. In relative terms, this was a considerable punishment, but the offenses for which he was prosecuted seemed disconnected from the events on the ground. "Killing wildlife without a permit" doesn't quite capture the act of feeding poison to hyenas. "Creating a disturbance" doesn't adequately reflect the connection that people drew between the hyena killings and the hyena attacks. It was not that I thought the farmer should have been punished more severely. He seemed as much a victim as everyone else involved. It was more that the law seemed vague and removed from the reality of the killings.

Following Muqtar's directions, we went to the place where the first child was attacked. It was an older part of town, the tin roofs rustier, the walls stained at the bottom by rain and pockmarked where the daub had lost its grip. Pointing to a house with a rusty corrugated iron fence, Simmey said, "This is the house of the child. The one who was eaten." I followed him through the gate into the compound, where we were greeted by a full-figured woman wearing a head scarf, her dress tucked in at the waist. There was nothing really to distinguish her from other Oromo women save that she had one less child because of hyenas. Simmey had already begun talking with the woman and I was afraid he was pressuring her, so I interrupted and asked through Arif if we could have a few words with her about her daughter. She agreed to that. I asked her daughter's name. I was anticipating that a crowd of people would soon be gathering in the compound, forcing a hasty departure, so the least I could do was to try to give her child's death some resonance. "Sumiya Yusuf." I wrote the name in my notebook but have never had to refer to the notes since. She committed herself to my memory. Sumiya's mother took us to the back of the house and showed us the place where she had been playing at the time of the attack. It was at the back of an outbuilding, in a place unseen by anyone in the house. The hyena came through a hedge at the back and took Sumiya away without anyone's knowing. There was nothing more I could think of asking. By now a crowd was gathering at the gate and people were spilling into the yard, so I gave her my condolences and left with Simmey and Arif on my heels, both wondering why the rush. I wanted to get out of town as quickly as possible.

As recently as 1998, spotted hyenas were classified as "vermin" in Ethiopia. It's a powerful word, vermin, as it confers on all members of a species not only a paucity of value but a status as antagonistic to the greater human good. Vermin are pests; they bring disease, ruin crops, invade houses, and attack people. In places other than Harar, hyenas fill the role of vermin like seasoned professionals. They penetrate livestock enclosures, break down the doors and walls of huts, drag off sheep and goats, and children; they stagger, rabid, into villages, snapping at legs like maniacs, dripping with diseased saliva; they break down hedges to create their paths across the countryside and trample maize in their run-around games. Yet the third millennium saw the establishment of the Ethiopian Wildlife and Conservation Authority and blanket protection for Ethiopian wildlife. At the stroke of a prime minister's pen and in the classic tradition of conservation rhetoric, hyenas were transformed from "vermin" into "natural resources," so that killing one could only be done with the right permission and after having paid the right money. It costs a foreign national US\$180 to kill a hyena. Comparatively, this is not a large sum—killing a leopard costs \$4,600—but it still means protection and it implies that hyenas are valued, if only in terms of foreign revenue. Meanwhile, Ethiopian nationals are prohibited from killing hyenas unless they hold a license. Old farmer Elias would have to jump through a lot of bureaucratic hoops and employ a licensed hunter if he wanted to get rid of a local problem hyena. Too hard, too expensive, and good news if you happen to be a hyena.

The blanket protection of wildlife seems to have been a knee-jerk policy designed to halt the rapid depletion of wild species in the country. Rather than take the time (and expense) required to determine which species need protection— which are abundant and which are going to be valuable "resources"—why not protect everything, stop the rot, and decide later which species don't need to be on the list? From a global perspective, this is considered progressive policy. The top predators, the keystone species, are preserved when otherwise they might have disappeared.[7] These charismatic fauna attract tourists who bring foreign income, and the economic benefits trickle down to the population at large. But try selling that argument to a bereaved mother in an East Hararge backwater where the nearest thing to a tourist the town had seen was a skinny anthropologist whose contribution to the local economy totaled two dollars and fifty cents for some coffees at one of the cafés and a couple of bottles of water. When it comes to the costs and benefits of maintaining predators in the landscape, mutuality and meeting halfway is an absurd notion, as far removed from the reality on the ground as the tourists flying ten thousand meters overhead, their silver tails trailing behind as they streak toward Dubai. The locals pay as much as their lives, and the benefits tend to drift off to somewhere far away.

David Quammen called this the "muskrat conundrum," after the findings of small-mammal ecologist Paul Errington. In the wetlands of northern Iowa, Errington studied the effects of mink predation on muskrat populations. What he found was that healthy muskrats in established territories with good access to resources were largely inaccessible to minks. Meanwhile, the displaced and dispersed muskrats on the margins constituted the majority of mink prey. The overproduced young, the weak, and the homeless were what Errington called the "wastage parts" of the population, and these were the favorite prey of minks. Quammen's muskrat conundrum draws a parallel between mink predation on muskrats and the conservation of predators amid human populations. It is usually the dispossessed and marginalized who must bear the major costs of maintaining top predators in the environment. Meanwhile, the spiritual, aesthetic, and economic benefits are enjoyed by those of us "cozier" muskrats. The conundrum: how can these costs and benefits be more equitably distributed among the entire population?[8]

Riding home on the bumpy road from Kombolcha, I felt guilty and ashamed. I was fond of hyenas. I was spending my nights with a whole bunch of them, running around the hill playing chase, teaming up against dogs, ruffling fur, being bitten, and wandering around the Old Town looking for food. I felt guilt by association, as if my membership in the Sofi clan conferred some kind of responsibility for the death of a little girl in Kombolcha. On top of that, I felt guilty for having brought an intrusive sideshow to an already disrupted community. The farmer responsible for killing the hyenas would no doubt hear about the farenji who came to the village asking questions and snooping around. He'd also be told of the local man who offered to take me to see him, information that I imagined would be not a little unsettling. The villagers who had killed the hyena that morning and tried to burn the body would have to make a hasty burial and wonder about the ramifications of what they'd done. The little girl with the swollen face had been made the center of attention in front of a strange white person for reasons she didn't understand. And Sumiya Yusuf's mother, rather than quietly sharing her grief, had had it laid out in front of her by a farenji and a crowd of locals. Even diverting the local administrators from their regular work seemed to me like an unnecessary and arrogant intrusion.

To cap it all off, I felt guilt by association with the global forces that I represented and that made these people so powerless in the face of dangerous hyenas. Blanket protection tied their hands in the service of an international, postcolonial master known as wildlife conservation. The master served up the ethos, rationale, and expectations behind Ethiopia's wildlife conservation policies. Proclamation No. 575/2008, which established Ethiopia's Wildlife and Conservation Authority, is replete with the rhetoric that fosters conservation of "wildlife resources" for their "sustainable use," in line with the "global development utilization and conservation standard."[9] The authority has policies and strategies for wildlife conservation based on international principles and conventions. The dangling carrot for the authority to adopt these is the revenue derived from wildlife resource–based industries such as international trade and tourism. The goal: to be ranked in the top five wildlife tourism destinations in Africa by the year 2020. The means: arrest the decline in wildlife populations, preserve habitat, and implement community participation in plans based on international principles. The implicit but unwritten truth, however, is that community participation will at times involve being food for the resources.

Like my presence in Kombolcha, the rhetoric didn't sit well with the established hyena/human relations in the town. The policy of arresting the decline of wildlife by means of protection was never going to create a wildlife haven around Kombolcha. The forest was almost entirely used up for firewood, and everywhere khat grew in its place. The only animal species that could persist were the odd crop raiders, like monkeys, gazellelike dik-diks, and those who could live off what the human population left out for them to eat: hyenas, mongooses, rats. Protection of wildlife in Kombolcha served largely to arrest farmers who exacted retribution on barn-raiding hyenas. But more than that, this strange language that considered hyenas "resources" was an awkward square peg for Kombolcha's round hole. The wereda leader's talk of boundaries that must not be crossed and locals' fears of hyena retribution were at odds with the rhetoric of wildlife resource management and sustainable use. After all, you don't discuss sending a person to negotiate a truce with a resource. On top of Quammen's muskrat conundrum of how to more evenly distribute the costs and benefits of maintaining top predators in the environment, there is another: how are we to get the locals to conceive of wild animals in the same way that we do? And are we justified in our efforts to do so?

# IO

## BLOOD OF THE HYENA

Back in Harar, I was making good progress following hyenas around Jugol. At the same time, though, I was experiencing difficulties where I lived. My landlady's partner, Faysal, was fond of khat. On arriving home from his farm midafternoon, he set himself up on a platform in the living room with a supply of water and khat, and there he remained until late at night. By about 10:00 P.M., he'd chewed enough khat to make him sufficiently paranoid to imagine a host of thieves, murderers, jinn, and hyenas gathering outside the compound, looking for an opportunity to get in. The lock on the compound gate required a key from both inside and outside, but this wasn't enough to assuage Faysal's fears. Perhaps hyenas could pick locks. So he slid a bolt into a recess in the wall, preventing access from outside even to me with my key. Hence, after following a hyena late into the night, I often arrived home to find the gate bolted. The wall was scalable, but the shards of glass embedded in the ridge and the three lines of barbed wire made it difficult to climb over. I could call Faysal, whereupon he woke up and opened the gate for me, but on nights when the phone network was down, which were many, I found myself stranded in the lanes of Jugol and had to resort to waking the clerk at the Ras Hotel and asking for a room. Neither the clerk nor I was happy with that arrangement. I needed alternative accommodations.

When I was due to meet the state president, Nureddin suggested that I take the opportunity to ask for an office in Jugol. If I could find the right place with twenty-four-hour access, it would be much easier to come and go while following hyenas. Not only that, but such an office would provide a comfortable setting for conducting interviews with locals. When I raised the matter with the president, he was sympathetic but said that this was an issue for the municipality to resolve. Nevertheless, he passed on a message to the mayor, asking that I get some assistance. When I later met with the mayor, he supported the idea and suggested a few places where I could set up shop. But after needling through the details, none was entirely suitable. Nureddin suggested that I could take over an office at the police station by the main market, but I was on good terms with the police and didn't want to ruffle any feathers by having someone evicted from their office. Finally, we arrived at a potential location. One of the officials in the Department of Land owned a vacant shop next to the main market. It was central, it had twenty-four-hour street access, it was secure, and it was near the police station.

The owner of the shop was a man named Hashim Idris. I knew him well from burcha at both Nureddin's house and the house of the former mayor, Zeyadan. Short of stature but powerfully built, Hashim had a deep, intimidating scar that ran across the bridge of his nose and through one eyebrow. Nevertheless, he had a kindly face. I met with Hashim at his shop, which fronted on the metaled road leading down to the main market. The position was ideal. In fact, it was just outside the compound gate that Kamareya and I had encountered the Aboker hyenas. The building was old and constructed of stone and mud, in the traditional style of Harari houses. The doors had no locks; according to Nureddin, Hashim's reputation was such that no one in Harar dared to enter without permission. The interior was cluttered with boxes and sewing tables. Prior to working for the government, Hashim had been a tailor, and this was his shop; his residence stood directly above the shop, so that both dwellings shared the same compound and gate. Hashim promised to empty out the room, install some locks, give the place a coat of paint, and put in some mats, a desk, and a chair. He also convinced the mayor that the municipality should cover the rent. All I had to do was have the shop appraised by someone from the municipality, agree on a rental price, and have a contract drawn up. It all sounded so easy, but I was wary and pessimistic. I feared getting caught up in the cogs of the bureaucratic machine.

After I had the office appraised, I was directed to the Department of Land's legal office to have a contract drawn up. I climbed the precarious, disintegrating wooden stairs of the municipality building and passed by the crowds of people waiting in the halls. Haile Selassie was long gone but the tradition of petitioning persisted; scores of people gathered at government offices to impress upon ministers and directors their personal plights. My appointment was with a mere legal officer, so I had no need to join the queues. I walked right into the legal office, ready to cross the final bridge between my nights of homelessness and a secure base of operations.

It would be a gross exaggeration to say that the man I found in the legal office greeted me. He was writing as I entered, and without looking up or dropping a word, he nodded almost imperceptibly in the direction of an empty chair beside his desk. I sat obediently. He was as thin as I was, with the tan skin of an Ethiopian who spent his days indoors. His close-cropped hair revealed a prominent brow over round eyes and a thin nose, suggesting that he was Amhara. A pair of narrow-rimmed glasses were perched precariously at the end of his nose. Without looking up at me, he fished through a stack of manila folders to find my file. "Your name?" he asked, as he scanned the document. I spelled it out for him. "The office is where?" I told him the location and he checked it against the file. "For what purpose are you using the office?"

The question almost pitched me off my chair. It was not unexpected—I knew I had to give the details for the contract—but from the stern mouth of the legal officer, it sounded like a challenge to justify my very presence in Harar. What was my purpose? Here I was, in the legal office of the Department of Land, asking for an office that would make it easier for me to follow hyenas around and watch them rummage through bags of garbage! Some of the people in the hall were waiting to plead with some director or other to let them remain on their land rather than be resettled, and I was applying for a place to nap after traipsing up drainage lanes on the tail of a hyena. Who cared about the hyenas, anyway? Nothing was changing; the hyenas were still coming and going every night and the bones were still being eaten. None of that would change as a result of my research. I couldn't think of a single thing to make my research in Harar sound worthwhile, let alone justify my being granted an office from which to do my frivolous work. Suddenly bereft of confidence, I stammered out what I was doing in Harar and why I needed an office, and then sank back in my chair, waiting for the ubiquitous bureaucratic "No," which I was beginning to feel I deserved. After hearing me out, the legal officer stopped writing. He put down his pen, sat up in his chair, and brought his hands together. He looked me in the eye for the first time since I'd entered his office and said gravely, "This is very important research."

I had my office. Hashim made the changes he promised, I added some cushions for sleeping on, taped a map of Harar to the wall, and my hyena research center was established. The shop front had no glass, just a steel door and a tiny wooden window. The back had a door that opened onto the compound. By day, I conducted interviews in private or typed away at my desk while Hashim's kids watched me through the curtain that hung across the back doorway. At night, I went to the Suqutatberi hyena hole and waited, usually in vain, for a hyena to enter Jugol. When I lost the tail of a hyena or felt I'd done enough waiting for the night, I returned to my office without disturbing anyone, except perhaps the indigent who slept in the adjacent doorway. I slept in my office until about 7:00 A.M., at which time I went home, had breakfast, and slept some more before beginning my day's work with the human population. By day, I was gey woldi, "child of Harar." I met with friends for coffee, bought khat at Assumberi, and had burcha at one of my favorite burcha houses. By night, I was gey waraba. I emerged with the setting sun and made my way to the hyena-feeding place. After several hours of watching and waiting, I ventured into Harar with a hyena companion and traversed the drainage lanes, looking for bones, avoiding people, and altercating with dogs. My double life took its toll on me physically; my teeth hurt from chewing khat, I was always tired from the late nights, always ill from respiratory or intestinal ailments, and covered in flea bites from sitting on the hill with the hyenas. But it was a small price to pay for the progress I was making.

My first successful following of Willi began in the center of the Old Town rather than at one of the hyena holes. I'd left the feeding place at 9:00 P.M., when the feeding was over and the hyenas were settling down for an hour's rest before attempting to enter Jugol. Professional anthropologist that I was, I went to pass the time with a beer at the Wesen Seged Hotel on the north side of Feres Magala.[1] I sat in a dark corner of the veranda, watching the last of the khat sellers pack up and go home as the city wound down for the night. At about 10:00 P.M. I emerged from my hiding place and headed toward the Suqutatberi hyena holes to wait for hyenas entering into the Old Town. But I barely made it past the main market before I encountered a hyena in one of the lanes. The hyena looked up and must have recognized me, because he shot me only the briefest of glances before continuing to chew on the bone that he'd extracted from a bag of bones and horns. It's uncanny the way hyenas can recognize individual hyenas and humans so easily in darkness and at a distance. I moved closer, shone my flashlight, and discovered that it was Jalla, a youngster of about the same age as Baby and Kamareya. Jalla picked up a horn and carried it right past me to the top of the lane. I followed and found another hyena standing at the top. The second hyena recognized me as well and I shone my flashlight, asking out loud, "Who are you?" I took a closer look and saw that it was Willi.

We were both taken aback by this chance meeting in a dark lane in the middle of Jugol. The context was strange. Previously, we had always met up at the feeding place. Either I found Willi among the Sofi hyenas when I went out to the hill, or he found me sitting at the shrine. From there we went and explored the road or wandered down to the garbage dump. But in the lane there was no reference point, just our physical selves removed from the usual context. This is why I had to take a good look at Willi before I recognized him. Normally, there were other clues to his identity, such as his favorite place to lie down, or his audacity in front of (or behind the backs of) tourists. Even the behavior of other hyenas toward Willi gave an indication of who he was. At the time, the higher-ranking females had ceased indulging him for his cubbish ignorance of the social hierarchy; Dibbey in particular was teaching him his place. If I saw Dibbey displacing or even harassing a young hyena, it was a fair guess that the victim was Willi. When I encountered the decontextualized Willi in the lane, I realized how much his identity—in my eyes—was interconnected with the places he favored and the hyenas and people with whom he associated.

For his part, Willi looked as disconcerted as I. He looked at me wide-eyed, head and shoulders lowered, forelegs tensed, his neck angled to one side as if ready to turn and run at the first hint that something was amiss. Rather than amble directly over to me and attempt to bite my arm, he simply stood at the top of the lane and stared. His misgivings were understandable. I was wearing my usual jacket, hat, and boots, so the physical me was familiar, but there in the lane in the middle of Jugol, the contextual me was altogether new. The standoff lasted only a moment, though. I switched off my flashlight and spoke in a soft voice while Willi dipped and raised his head, then made an almost imperceptible move toward me. He stood and waited while I approached. Jalla went off in the direction of the main market while Willi swished his tail and went the other way. I caught up and walked beside Willi.

We arrived together at the bottom of a channel lane. These lanes permit the passage of people but are bisected by narrow channels that carry runoff and wastewater to the drainage lanes proper. Halfway up the channel lane we found a bone on which Willi crunched for a minute before we looped around and returned to where we'd just met. Jalla returned and, with Willi, resumed munching on the bones and horns that by this time were scattered across the lane. At some imperceptible signal, they both stopped to look to the top of the lane, where an unfamiliar hyena appeared. On seeing me, this hyena uttered a low-pitched croaking alarm call and scarpered off up the lane. Jalla decided to follow, and we went to the intersection, where I could see the stranger waiting up. Jalla joined her and they went off together to the main market, while I turned and went back to wait for Willi. When I looked beyond Willi, there was yet another hyena, this time at the bottom of the lane. Hyenas seemed to be coming out of the cracks in the walls! This stranger took a look at me and went back in the direction from which she'd come, turning toward Suqutatberi Road. Willi followed with me close behind. It was at times like these that I felt not at all like a researcher in control of his situation but instead like a naïve hyena dependent on the guidance of others. Not unlike Willi.

Before we arrived at the stepped lane that descended to Suqutatberi Road, we passed a mound of rags butted up against the corner where two walls met. I knew that there must be a human underneath and so too, no doubt, did Willi. Flesh and bone beneath the material, and not fifty meters from where Willi and Jalla had been eking out the tiniest morsels of flesh from discarded horns and bones. Yet Willi and the unknown hyena passed by the homeless person without even stopping to sniff. The Amhara author Nega Mezlekia describes how this passing up of potential food would be unheard of in Jijiga, the town one hundred kilometers east of Harar.[2] Mezlekia's authoritarian schoolteacher, Mr. Alula, avowed the benefits of hyenas in a town where anyone alone in the streets after dark would be devoured, skin, bone, and clothing. The Jijiga homeless preferred hunger to predation and pooled their resources to rent rooms with secure latches. Accordingly, Mr. Alula considered large numbers of homeless people vividly indicative of a "shortage of hyenas." In Jugol, the rule didn't appear to apply. We passed mound after mound of sleeping humans without so much as a sideways sniff. The taking of a baby from his homeless mother outside Selassie Church was more than a tragic and pitiable loss of life. It was an exception.

Once across Suqutatberi Road, Willi found a drainage lane. This one had concrete slabs on top of the drain, so it was a little easier for me to follow. On a grassy verge about halfway down, Willi noticed a feathery lump: a chicken carcass. When I thought he was going to take a bite, he instead pressed the side of his muzzle into the chicken and turned his head, as though wiping the side of his mouth with the feathers. The musty smell must have appealed, because Willi followed his nose with the rest of his body and pushed forward so that he lay across the chicken. He folded his legs and twisted his body so that the carcass disappeared completely beneath him, then made another roll until I could see a protruding claw. Willi stopped midroll with all four of his paws in the air and his belly exposed to me. Trusting. He gave me a quick glance and then did a little wiggle dance with his back. I never knew that rotting carcasses could be so pleasurable. Then, midroll, Willi suddenly snapped out of his reverie and stood up facing southeast. I held my breath and listened, trying to make out what he was attending to. The faintest of oowhoops echoed across Mount Hakim, and Willi froze, his ears facing forward, listening to the information. No wonder people in Harar thought hyenas' oowhoops carried important messages. Willi held still until the hyena finished his bout with a long, mournful ooooo and then made off down the lane, leaving the chicken behind for the next hyena to anoint herself.

We went all the way down to the town wall near Abadir and then followed a lane along the inside of the wall, back across Suqutatberi Road and toward the north. Just to see if I could, I passed Willi in some places and took the lead. He happily followed each time, but halfway to Argobberi he couldn't resist biting my leg. We ascended the lane that led past the shrine of Sheikh Hashim, and we both got a start from two dogs who materialized at the top of a compound wall and began barking like maniacs. I directed my flashlight at the dogs, but Willi was reticent about walking beneath their vantage point. He went off to the left and I followed, leaving the dogs behind still barking their lungs out.

The relationship between hyenas and dogs in Harar is incredibly complex, although there is one rule that applies in most cases: hyenas prefer to mind their own business, while dogs want to cause trouble. There are exceptions. The dog without a master to impress or property to defend will keep quiet when he needs to pass by a group of hyenas. He wedges his tail between his legs and hurries past without a sideways glance. However, when dogs are in packs, with their masters, or defending a compound, they turn maniacal. They have an overwhelming compulsion to bark at hyenas and run after them. For an anthropologist studying hyenas, this is both annoying and convenient. When you're following a hyena around Jugol, it frays your nerves when, at random times and places, the still of night explodes into maniacal barking right beside you. With experience, though, you learn where the maniacs live and come to expect the barking when you pass by their compounds. It's also frustrating when a hyena you're following is chased off at speed by a pack of dogs and you lose him. On the plus side, however, the barking is a dead giveaway that a hyena is present. So when you lose the tail of a hyena you're following, you simply stop and listen for the barking maniacs, then head in that direction.

Hyenas usually react to dogs with deference. When a dog comes running toward a hyena barking its chops off, the hyena will normally canter off to a safe distance. If the dog is accompanied by others, the hyena might have to run. But they seem to take this in stride. At the playing field, I watched once as a pack of dogs went running toward a mother and cub who were lying down together in the field. The dogs singled out the cub, who got up and ran around his mother while the chasing pack nipped at his heels. Meanwhile, his mother lay on the ground, motionless apart from her ears, which swiveled to follow the action. Sometimes, however, hyenas do fight back. I occasionally heard dogs going crazy on the track beside the feeding place, followed by the sound of a hyena growling. At that, several hyenas who were feeding abandoned Yusuf and the bucket of scraps and went racing down to the track to engage the dogs and run them off. I have even seen three hyenas pursue a dog across open country as if they were hunting. Baby in particular had it in for dogs. Whenever she heard an altercation, she ran to join in, or if dogs passed along the road below the hill, she went to the edge of the hill and growled at the four-legged passersby.

Willi and I emerged from a lane at the road that passed through the gate at Argobberi. It was late by then, about 1:00 A.M., and the khat sellers had all gone home. There were only a couple of indigents sleeping under shop awnings, and a dog lying asleep in the middle of the road. Willi ducked off the road, presumably to avoid the dog, and went through the covered opening beside the old Argobberi gate. He headed up the road toward the shrine of Aw Warika at the top of the hill. This was interesting; we were outside Jugol and heading in the direction of Aboker hyena territory. I was interested to see how far Willi would go toward the Aboker dens, which were just over the hill and beyond the slaughter yards. There was a truck backed up to one of the yards from which cattle were emerging and plodding naïvely to their bloody fate. Willi was interested in the truckload of cattle, but he stopped halfway up the hill. I walked past and encouraged him but he refused to continue, instead investigating a drainage ditch beside the road. He jumped out of the ditch and went back in the direction of the Argobberi gate, where we found a big orange dumpster full of garbage. Willi jumped in and foraged while I waited alongside.

When Willi exited the dumpster and headed toward the feeding place at the bottom of the hill, the dog who had been lying in the middle of the road came racing toward us, barking madly. Willi ran off with the dog nipping at his heels, and the chase took them toward the green zone at Argobberi, a patch of ground beside the wall that was being rehabilitated and beautified. I was incensed and went running after the dog, who, looking over his shoulder, soon realized that he was being pursued. The dog ceased his barking and veered off, running across the road while I sprinted after him. He jumped from the road onto the deserted common and made off past a couple of houses in the direction of Gunda Fera, the suburb that was once a leper colony. I pulled up at the edge of the road and glared at the dog, breathing heavily from the exertion. That was when I noticed Willi standing right beside me, his tail stiff as an antenna, glaring at the dog. He'd turned about when the dog gave up the chase and joined me in turning the tables on his pursuer.

I was surprised. Normally, a hyena's immediate reaction to the stamping feet of a human is to run off. When Yusuf's youngest child, Ajebbo, arrived home from the market at night, she ran between the shrine and the fence, stamping her feet as she descended the steps. At the sound of stomping feet, the hyenas scattered, even though it was a regular occurrence. Running humans simply signaled danger, and hyenas invariably ran off; it was a reflex action. But Willi had done just the opposite, turning toward me when I went running after the dog and joining me in the chase. In midflight he'd summed up what was happening, discerning my intention and the dog's reaction. He wheeled about and with characteristic hyena fleetness caught up with me as I reached the edge of the road, just in time to watch the dog disappear into the night. He trusted me, guessed rightly my intentions, and backed me up. But before I had time to absorb all this, and with typical hyena pragmatism, Willi turned and wandered past the mosque toward the cemetery. I stayed where I was; it was unwise to wander off into the farmland, where farmers shot at khat thieves in the darkness.

Election time in Ethiopia came around with warnings from the Australian consulate. The previous elections, in 2005, had seen protests in Addis during which people were shot, so I was advised to stay away from public gatherings and lie low on Election Day. I didn't pay much attention. I was really only concerned with the elections because of Nureddin. He was running as an opposition candidate and complaining about dirty tricks by the ruling party in Harar. Otherwise, my focus was on the hyenas and following them around Jugol.

On the eve of the election, I watched the hyena feeding, as always, and then waited outside the hyena hole for hyenas to go into Jugol. Willi led me in for a quick scan of the common inside the hyena hole but soon exited and returned to the feeding place. I chose to remain on the road to wait for another hyena and was soon rewarded. Kamareya came ambling along the road, took a sideways look at me, and entered Jugol.

We followed the lane that ran parallel with the wall in the direction of Suqutatberi. The first food that Kamareya found consisted of some unidentifiable scraps at the edge of a drain. He was joined by Baby and they ate together peacefully until some people came from the top of the lane. On hearing their approach, Kamareya took off first. I followed him to a low arch over a drain, where he hid while the people passed by. He looked at me as if waiting for the all clear, which I eventually gave. At that, Kamareya went straight for the entrance to the series of drainage lanes that led to the main market. There was an arch at the entrance that hyenas easily slipped under but that would have forced me to go through on hands and knees, so I ran around to where I knew the lane exited. I arrived and waited for Kamareya, but the first hyena to emerge was Baby. I hadn't yet followed Baby successfully in Jugol, so I left Kamareya to his own devices and followed on Baby's tail as she headed toward the main market.

The first thing I noticed about Baby was her speed. Willi was a bit of a slowpoke and relatively easy to keep up with; Kamareya was a bit faster, and I usually broke into a sweat following him; but Baby was positively mercurial. I had to halfwalk, half-skip as I followed her, yet without breaking into a run, which would frighten her into running off. Yet all the while she never broke into so much as a trot—such is the pace at which hyenas are capable of walking.[3] She only slowed down a couple of times, once when she stopped to greet another hyena at the top of a drainage lane, and once when she had to run to get away from some people. She stopped, turned to look at me, and waited for me to catch up. Once I arrived, though, she headed off again at a quick pace. I felt rushed, like I had a tenuous grip on Baby's tail, that if I eased up even a little, Baby would disappear. This was not a good frame of mind, considering what happened next.

When someone appeared in a lane ahead of us, Baby took a detour along a series of lanes that looped around and brought us out on the other side. She marched on toward the main market and broke into a trot when she perceived someone coming along a lane to the right. The person said, "farenjo, hyena," as he passed me by, and I answered, "indeed," as I hurried to catch up. Baby slowed to allow me to get closer, but once I was close enough for her liking, she quickened her pace again and turned left just before the police station to descend a set of steps. I noticed a policeman at the top of the steps as I turned to follow Baby, who was disappearing around a corner at the bottom. As I hurried along, I heard from behind me, "Stop!" Thinking it was one of the local policemen, I responded in Harari, "No problem, I'm with the hyena."

As I continued down the steps, the voice again called out "Stop!" Rather than lose Baby, I turned and ran back toward the policeman so he could see that it was me and I could get back to following Baby before she was gone. By the time I was five steps from the policeman, I realized that he was pointing his automatic rifle at my head and had a most terrified look on his very youthful face. I knew then that I wasn't going to be following Baby any further that night.

Once I'd stopped, I did something that in hindsight was ill advised, or, less charitably, stupid. I began fishing in my jacket for my magic letter from the director of tourism, sport, and culture. I still hoped that the policeman, given one look at the letter, would lower his gun and let me go racing after Baby. It was only when I saw his terrified expression that I realized the foolishness of what I was doing. I slowly removed my hands from my jacket and held them in plain view, and I obeyed when he indicated that I should sit on the chair at the entrance to the police station. Once in the light, I noticed the blue-gray camouflage pattern of the policeman's uniform, and it dawned on me what was happening. The local Harari police, who knew me, had been replaced by federal police in anticipation of the election. The policeman detaining me had no idea who I was or why I was following a hyena, nor did he understand Harari. He was Amhara. No wonder my explanation had fallen on deaf ears. In the end, the young policeman summoned his commander from the station and I was questioned. I dropped as many names of police commanders and media liaisons as I knew, and before I could even call someone to come and vouch for me, I was sent on my way. This left me wondering why all the fuss in the first place. Rather than go looking for Baby, I thought it prudent to go home.[4]

That was not the only time I was threatened by someone in Jugol. On another occasion, I followed Willi around the Suqutatberi quarter and emerged at the Argobberi gate. Willi headed down the road toward the farmland, so I turned and headed back through the gate and along a lane that would lead me to my office. Up ahead, a man sitting on a step stood up with a terrified look and asked what I wanted. I said in Harari that I was on my way home. Either this failed to allay his fears or else, coming from my mouth, the Harari word for "home" sounds like Harari for "to kill you!"—because the man picked up two rocks the size of grapefruits and held them ready to throw. By this time I was only a few meters away from him and could see the look of terror on his face. I thought that if I turned around he'd throw the rocks for sure, so I edged past him as best I could in the narrow lane and ran with my head down. The rocks went whizzing over my shoulders and bounced down the lane ahead of me. I stopped, turned, and real-ized that my night-vision scope had dropped out of my jacket pocket. The man came toward me with it and put it on the ground for me to collect. At the same time, his friend came out of the compound opposite to see what was happening and took him inside. I called out after them in Harari, "Why!"

The answer: Ethiopians fear crocutaphiles and theriomorphs—people who associate with, or else transform themselves into, hyenas. For Amhara people like the young police officer, these kinds of people are known as *buda*. By day, buda are humans, usually lower-caste people such as blacksmiths, tanners, and basket makers. As with lower castes elsewhere in the world, these people are at once denigrated and feared. They are denigrated for their landless status and feared for their jealousy of the landed classes, a jealousy that is manifested in the buda's gaze. A glance from a buda can bring on misfortune, wasting disease, sterility, and all manner of ills. Hence an Amhara woman carrying a bottle of milk will conceal it in a bag so as not to attract the covetous eye of a buda. When an Amhara person is given a present, the person will politely put it to one side, to be opened later when there is no one around who might cast a jealous eye over the contents. Anyone could potentially be a buda, so anything that might elicit jealousy must be con-cealed. Moreover, and more to the point, buda are closely associated with hyenas. They ride hyenas at night, seeking victims to devour, or their hyena familiars bring them human corpses. More than that, though, buda can transform themselves into hyenas and exhume corpses, which they reanimate and bring home to use as household servants. This is why the hyena-skin armband that Faysal told me about protected its wearer from buda—a case of like curing like. According to Ronald

Reminick, who did his fieldwork among the Manze Amhara of central Shewa, the buda were hard to distinguish. They could be thinner than normal people (because their blood was supposedly thinner) and they could have lighter skin. These characteristics alone are not enough to identify someone as a buda, but a skinny white person in the company of a hyena at night in Jugol would have a high probability of being a buda.

Among the Hararis, there is no lower caste (although other ethnic groups are considered inferior). All Harari women make, or should be able to make, baskets, and all Harari men are able to manufacture things without being denigrated. In his 1974 doctoral dissertation on social organization in Harar, Sidney Waldron referred to a class of untouchables in Harar known as the *gey garach*, or "family of the city."[5] These people were apparently feared for their possession of the evil eye, but they were not an artisan class; they were a lineage. By the time of my fieldwork, these people were no longer distinguished, or else nobody wanted to mention that they were. Still, many Hararis feared jealous buda in the vein of the Amhara tradition. They concealed or disguised valuable items and opened presents in private. They also feared those who could transform themselves into hyenas at night and go out seeking victims. In particular, the Argobba and the Issa Somali were believed to be capable of this. They chewed on a certain kind of wood before lying down and undergoing the physical transformation. What they did afterward was not as clear, but presumably they captured and devoured humans. My translator friend Arif told me of an incident in which a child from an Argobba village was taken by a hyena. When some soldiers tracked and killed the hyena, they found that he had human hands and feet instead of paws, and a human penis. For a Harari man under the influence of paranoia-inducing khat, the possibility of encountering a human/hyena familiar in the dead of night is very real.[6]

Beyond the problems associated with fearful people, there was one environmental element that was both problematic for me and a solace: power outages. The most reliable thing about the power supply in Harar was that it was guaranteed to cease without warning for at least eight hours a day. Sometimes the power remained out for up to twenty-four hours, and when it finally returned, cheers rang out across the town. Sitting on the hill with the hyenas on moonless nights, I could normally see by the dim glow of the scattered yellow streetlights in Jugol and the dim globe that hung outside Yusuf's house. While it was hard to identify individual hyenas by this light, I could at least see where they were. But when the power went out on a moonless or cloudy night: blackness. I found myself sitting on a hill surrounded by hyenas, unable to see what any of them was doing. This was especially unnerving when Willi was in the mood for biting.

Still, when I was in the company of hyenas in Jugol, the power outages made the place feel safe. All of a sudden, we could walk the streets unseen by people, even people only a few meters away. It brought a feeling of security and freedom from persecution. The dogs still barked, but there was no light to betray our presence to people. For all they knew, the dogs were barking at ghosts. "You have *waraba dam*," said Abdulmuheimen when I told him about my late nights. "Blood of the hyena. A man who stays up late chewing khat, extending his burcha into the morning, this man has waraba dam. You stay late at night with warabas, so you have waraba dam." He was disquietingly close to the mark. I was emerging from my office with the sunset and spending my evenings at the hyena-feeding place. After the feeding, I lay with the hyenas on the hill for a couple of hours, picking up fleas and listening to the noise of humans subside before getting up and wandering into the Old Town with hyenas, looking for food. I was disturbed by those irritating dogs and felt threatened by the people who were out in the lanes late at night. Ironically, I felt safest when hopping along in the drainage lanes, enveloped in the darkness of a power outage and in the company of deadly, dangerous hyenas. Perhaps the rock-throwing khat addict's suspicions were justified after all.

# II

## ACROSS A HUMAN/HYENA BOUNDARY

The night of June 11 began like any other night in Harar. By 9:00 P.M. the hyenas were settled and waiting for the human traffic to die down before they entered Jugol. Spotted bodies lay scattered about the hill. I got up and went to Yusuf's compound to say good night to everyone before walking up the steep road to the Wesen Seged for a beer and a hiding place. I expected I should be back by 10:30 P.M. to catch the hyenas passing through the hyena holes. But Yusuf had guests—family from just north of Harar—and he invited me in to spend some time socializing. I sat and fumbled a conversation in my hack Oromo before saying my good-byes and walking out into the night. The hyenas swiveled a few ears to follow my footsteps as I walked past the shrine and up the cobblestone road toward Argobberi.

As I walked, I could see at the top of the hill, in the middle of the road in front of Aw Abdul Mosque, a circle of people. On getting closer, I realized that there was a creature in their midst, lying on the road. A juvenile hyena. This was a busy stretch of road where trucks and minibuses rumbled through on their way to the Erer Valley, so I assumed that the hyena must have been hit by a vehicle. He was still alive but apparently in some distress.[1] He had just defecated, and he squealed as his body was intermittently wracked with convulsions. Meanwhile, a sheikh stood over the wretched hyena reciting verses of the Quran. The way the hyena writhed spasmodically as the sheikh murmured verses made it seem like his recitations were controlling some kind of possessive spirit who had control of the hyena. But the writhing was from pain and the verses were intended to cure, or at least alleviate.

I asked a man in the crowd if the hyena had been hit by a car. He told me that it must have been poison. That accounted for the spasms.[2] The hyena's whole body contracted and contorted; his head twisted sideways and he whined and squealed before his body untwisted and he lay on his side panting, as if trying to catch his breath before the next contortions inexorably came. His suffering was intense. An Oromo farmer with a muncha stood nearby. I asked why he didn't kill the hyena to end his suffering, adding, "You guys think nothing of killing a hyena out there in the countryside." He didn't understand English. Some girls giggled, whether nervous or amused I knew not which. I didn't know what to do; I found myself hoping the sheikh's recitations would work.

Before long, Yusuf arrived. I hadn't noticed, but someone must have run down the hill to fetch him. Yusuf turned the hyena over with his foot and asked me, "Is it Willi?" I said no. I didn't recognize this youngster as one of the Sofi hyenas, so I figured he must be from the Aboker clan. Yusuf called to a man in the crowd and asked for some matches. The man handed over a box from which Yusuf took a match. He struck it on its side, it flared, and then a deft wave of Yusuf's hand extinguished the flame. He held the smoldering match below the hyena's flaring nostrils. I'd seen something like this in the streets of Dire Dawa when a man suffered an epileptic seizure. Someone lit a match and extinguished it so that the man would inhale the smoke. I think it alleviated the helplessness of bystanders more than it did seizures. In the case of the suffering hyena, Yusuf must have figured that he was having a seizure like any human and tried to effect the standard cure.

I said I thought it was more likely poison that had caused the hyena's suffering. Yusuf concurred and gave instructions to a boy, who ran to one of the traders lining the road and returned with some limes. Yusuf took one of the limes and bit it in half. He knelt down and held the hyena's mouth in one hand while squeezing the lime with the other. In the back of my mind I considered the possibility that the convulsing hyena might be rabid, and I was concerned that he might inadvertently bite Yusuf's hand. Yusuf had difficulty holding the hyena's mouth, so he gave the lime to a man standing nearby and instructed him to squeeze the lime juice while Yusuf held the mouth open with both hands. I had heard about this treatment. When Harari girls wanted to avoid marriages imposed upon them by their parents, they sometimes drank bleach—a cry for help, an attempted escape. At such times, the girls were made to drink lime juice; the acidity of the limes was intended to nullify the alkalinizing effects of the bleach. For the poisoned hyena, the lime juice was useless.

A young man arrived with a wheelbarrow he'd brought from Yusuf's house. Yusuf must have sent him to fetch it, intending to bring the hyena to his home. I wondered how this little hyena would be received there. We lifted the panting hyena and placed him in the wheelbarrow. He twisted and squealed as Yusuf wheeled him down the road with the sheikh following behind, still reciting the Quran. About halfway down, the wheelbarrow collapsed. It was already tenuously held together, and the weight of the hyena was too much. The barrow thunked onto the ground, the frame skewed off to one side, the wheel to the other, and the contorted hyena almost spilled out onto the road. I took up one corner of the barrow, Yusuf and two men took up the others, and we carried the hyena the rest of the way to the feeding place, grateful for something to do.

Before we even arrived at the feeding place, the Sofi hyenas had heard our approach. The squealing hyena roused them, and four hyenas were waiting for us in the feeding area, their manes and tails erect, their ears pricked forward, attending to the noises emanating from the barrow. We put the barrow down in the middle of the feeding area. I suggested that Yusuf should try giving milk to the sick hyena. It was all I knew to treat cases of poisoning, and at worst the poor hyena would have sweet milk for his last supper. Yusuf called out to Naime, who went into the house and emerged with a flask of milk, heated and bespeckled with bits of black, the burnt, fragrant wood with which Oromo people infuse their milk to give it a sweet fragrance. Yusuf held the hyena's mouth open and I poured, not imagining for a minute that it might work. The hyena retched and squealed as though oblivious to the people and the treatment being administered; the Sofi hyenas lowed, groaned, and whooped, announcing to their clan mates the hyena's presence.

Within minutes, a large crowd of Sofi hyenas surrounded Yusuf, myself, and the poisoned hyena. The other people present had retreated to the step of the shrine, hesitant to be even standing at ground level. I counted at least sixteen hyenas in the light of the feeding place. I shone my flashlight onto the hill and counted another sixteen pairs of glowing eyes bobbing up and down in the darkness. Yusuf lifted the hyena from the barrow and placed him on the ground like an offering to the Sofi hyenas. I looked at Willi. He was as worked up as the others but, like them, reluctant to come forward. Yusuf retreated to the shrine and climbed onto the step with the rest of the human onlookers. I wanted to prevent the Sofi hyenas from getting to the juvenile, so I remained with him in the feeding area. Foremost in my mind was Kruuk's account of hyenas eating the carcass of another hyena, and I feared that a Sofi hyena who ate this juvenile would suffer from ingesting the poison.[3] But Abbas insisted that I retreat to the shrine. He and Yusuf practically dragged me back, leaving the poisoned hyena to the will of the Sofi clan.

Willi approached first. I had my heart in my mouth when he bit into the poisoned hyena and dragged him back across an invisible line between the hyenas and the people that the others were reticent about crossing. Koti was extremely riled. She looked almost twice her size as she pushed Willi out of the way and picked the young hyena up in her mouth. Almost effortlessly, she marched off with the hyena in her jaws, followed by a gaggle of whooping, lowing clan mates funneling between the trees and into the night. I sat and listened to the din of hyenas out there in the farmland, cloaked in darkness, bloodcurdling whoops and growls ringing out the macabre reception for the young hyena we'd brought into their midst. After the sounds faded, I went directly home and lay awake until 3:00 A.M., unable to sleep, with recurring scenes in my head of a retching, panting, desperate hyena on a bed of cobblestones, and with guilt over my own inaction.

The following night, I went with trepidation to the Sofi feeding place. I'd gone out in the morning to inspect the farmland and determine what had happened to the young hyena. Abbas assured me that the milk had cured the victim, that he had seen him running about with the others, but I guessed this was just an expression of Abbas's faith in farenji medical knowledge. All I found was trampled grass and trails leading every which way. The signs on the ground bespoke chaos; there was no sign of the poisoned hyena. Sitting in front of the shrine that night, watching the moon rise over the cemetery, I imagined a dozen hyenas feeding on the carcass and wondered who they would be. Koti arrived early, so too Baby and Kamareya. One by one I noted the arrival of individuals and marked their names with relief in my notebook. But I felt the greatest relief at the arrival of Willi. It was a one-sided sort of reunion, though. I went over to him and rubbed his head vigorously, while he simply nipped at my arm and walked past. Hyenas are unsentimental at the best of times.

As far as I could tell, all of the Sofi hyenas were fine; I imagined they must have killed but not eaten the Aboker hyena. Perhaps they were put off by the scent of a foreigner. Relieved, I thought that this was the last of the incident, but I was wrong. I met with Hashim Idris the following day and told him about the poisoned hyena and the events of that night. "That's interesting," he said. "On the same night I saw from my window a big group of hyenas in the main market making a lot of noise and marching about. They were excited about something." The Aboker hyenas had been listening to everything that went on at the Sofi feeding place that night, and they weren't about to let it go. The little poisoned hyena's lying down on the road outside Argobberi was a catalyst for the reignition of a broader conflict that was woven into the landscape and even the history of the town.

Hyenas define the boundaries of their territories in several ways, most of which involve some form of scent marking associated with physical features of the landscape. One way they mark territory is to straddle long blades of grass along territorial boundaries and deposit a scented paste from their anal glands. The paste is a strong-smelling marker of individual identity and, by extension, of the clan to which that hyena person belongs.[4] Another way is scratching the ground. Hyenas paw at the ground along boundaries and deposit scent from their interdigital glands. These two methods are complemented by their latrines, accumulations of feces in strategic positions. The bright white feces are visual and olfactory markers that identify ownership of the territory.[5] In places like Ngorongoro and Maasai Mara, these behavioral and physical markers give clear information to hyenas (and hyena researchers) about the delineations of clan boundaries and land tenure. In Harar, this system doesn't work.

In Jugol there are more than three hundred narrow lanes and almost no open ground. This alone is not enough to prevent hyenas from marking territories; theoretically, they could scent mark along a series of intersections, which should prohibit intruders from other clans. The problem is: what to mark? Most of the lanes are bereft of vegetation because of the heavy foot traffic, while any grass that finds a secluded place to grow roots is constantly trimmed down to ground level by keen-eyed goats who have run out of khat leaves. There is simply nothing upon which hyenas can deposit scent by straddling. And as for scratching scent into the ground, the cobblestones and flagstones make that a pointless exercise. Before the cobblestones were laid, it is possible that hyenas scratched the ground and deposited scent, but systematically to define boundaries? Difficult. Much of the scent would have been masked or dispersed by the feet of the thousands of humans, goats, sheep, and donkeys who walk the lanes daily. As regards fecal latrines, no chance of that in Jugol. Women's cooperatives comb the lanes every morning, sweeping up the candy wrappers, cigarette packets, discarded khat leaves, and any hyena feces deposited during the night.[6] In short, there was no way for the Sofi and Aboker hyenas to be able to mark, maintain, and recognize territorial boundaries in Jugol.

This is why I never witnessed a clan war in Jugol. The Sofi and Aboker hyenas combing the lanes of the Old Town had no clear idea of where their territories ended and began. They did generally stay on their own sides of the main market, which was close to the center of town, but where there should have been a line between the main market and Argobberi, there was actually a tremendous overlap. This in turn led to some bizarre interactions between members of different clans. The encounter I described between Kamareya and the Aboker subadults was just one example. On another occasion, I followed a Sofi hyena from the gate at Suqutatberi to the main market, where she encountered and greeted two hyenas whom I then followed over the hill and into the Aboker hyenas' territory. I had witnessed a peaceful greeting between hyenas from different clans. On another occasion I was watching Jalla eat a bone well inside what I thought was the Aboker part of Jugol, when two Aboker hyenas approached. These three were outwardly unconcerned with one another and went their separate ways. There could be no territorial dispute because there was no way to establish who was in the wrong place. Instead, there was only tension, indifference, or even cordiality. But this state of affairs applied only inside Jugol. Outside the wall there was a strictly defined, aggressively maintained boundary, and it fell right outside the Argobberi gate.

Two nights after the poisoning incident, something was afoot at the Sofi feeding place. At about 10:15 P.M., I heard a lot of lowing and groaning coming from Yusuf's compound. I went to investigate and found a bunch of hyenas sniffing around and making a lot of noise, their manes and tails erect. They completely ignored me. Then some growling hyenas came past me, ran onto the road, and headed up the hill toward Argobberi. I followed at a walk, figuring they wouldn't go far. Sofi hyenas rarely ventured in that direction, for that way lay Aboker territory. By the time I was halfway up the hill, I could see the hyenas at the top. They turned and came running back down the hill, growling and lowing as they came. Koti and another ran past me and into the feeding area, while the other two stopped beside me and turned to look up the hill. The two with me were Dibbey and Jalla, both quite high-ranking hyenas. I heard a lot of commotion down in the feeding area and was about to go and investigate when Koti came running back up the hill with another two hyenas. They ran past me, collecting Dibbey, Jalla, and me as they went, and we all went running up hill toward Argobberi and Aboker territory.

At the top of the hill I encountered a bizarre scene. The road was deserted of people. It was late in the evening, but at that time there would normally have been traders packing up, people taking the road home after visiting friends, and khat addicts making their sleepless rounds of the town. Looking along the side of the road, I realized that all of the people had vacated the road and were standing under the awnings of the shops that lined the roadside. Meanwhile, in front of the Argobberi gate ahead, I made out the silhouettes of six hyenas under the solitary street light. The Aboker hyenas had come to Argobberi! A man standing in front of a shop called out to me, "Hey, farenjo! Get off the road, hyena is dangerous animal!" But I ignored his advice. I felt pretty secure standing in the middle of the Sofi hyenas; in fact, I felt a kind of euphoric sense of belonging. My only concern was that the Aboker hyenas might attack, making the Sofi hyenas retreat to the feeding place at a run, leaving me trailing behind and vulnerable to the enemy. I made sure to stand well to the back of the others, who by this time were going crazy. The din was incredible as the two groups of hyenas growled and lunged at one another in a cloud of dust under the streetlight. But it was all ritualized violence, and not a single bite was inflicted. The hyenas charged at one another and made a big show of growling and posturing, until the Sofi hyenas turned and marched back down the hill whence they'd come, their tail standards raised in the fashion of a measured withdrawal. I remained on the road and watched as the Aboker hyenas turned and marched off toward their own territory, their erect black standards disappearing over the hill at Aw Warika. Each group had apparently made its point.

I should have seen it coming. In numerous conversations with people in Harar I'd been told about that place in front of the gate at Argobberi. Just a couple of weeks into my fieldwork, Adil told me about the frequent fighting there when Aboker females apparently came to steal males from the Sofi clan. Another Harari told me that a butcher often took meat and bones to the gate and incited the Aboker and Sofi hyenas to wage war. According to him, people present chose sides and cheered on their favorite hyenas. Indeed, so many people had told me about the Sofi and Aboker hyenas fighting outside Argobberi that I stopped writing their stories down in my notebook. It was just common knowledge, and I didn't want to miss anything else being said while making notes about something I'd heard over and over. This is folk ecology, usually taken with a pinch of salt when it comes from one of the locals. But when the stories mount up, they grow beyond anecdote into evidence, from which data come. Strangely, folk ecology remains folksy until tested by a (usually Western) researcher, but that's a subject for another discussion. What interests me here about the clan war at Argobberi, and the accounts and involvement of the local people, is how the significance of the place as boundary is a mutual construct of hyenas and humans.

Over and above the processes that led to humans' and hyenas' sharing the landscape around Harar, the events of the sixteenth century were integral to the making of that boundary. When Emir Nur ordered construction of the wall around Harar, he established the circumstances that would eventually prevent hyenas from maintaining territories inside Jugol. As Harar became surrounded by the Oromo, the town could only grow centripetally, becoming more internally complex. This made the town bereft of features on which hyenas could mark their boundaries and devoid of open spaces in which to engage in interclan conflict. But this is not to say that the place outside the Argobberi gate was the inevitable locus of interclan conflict for these hyenas. The boundary between their territories ran beyond the mosque at Aw Abdul and off into the countryside, where there was plenty of space to mark territories and engage in disputes. For the place to retain its significance, the hyenas had to keep the tradition alive; they had to engage in regular disputes and indoctrinate young hyenas as to its importance. For a young hyena like Jalla, the first clan war was a learning of the place, and that was reinforced by subsequent clan wars. It was through participation in ritualized and actual violence in that politicized, historicized, culturally constructed place that hyenas came to understand its significance. As a boundary, it was in fact a social construction.

And this is the interesting bit. The defined, contested, but agreed-upon boundary was not only a social construction of the hyenas; it was also a social construction of the human population. Through experiencing the place and listening to stories of hyena clan wars, the people of Harar came to learn its significance. They in turn passed these stories on to others. More than that, though, the people who witnessed the clan wars participated in the perpetuation of the site's significance to hyenas. Theoretically, they could have remained on the road and even chased the hyenas away. It would have made for an interesting test of nerves, though I suspect that the hyenas would have deferred. But, instead, the people vacated the road and gave it over to the hyenas, who in turn accepted the gift of open space for their disputes. It took the participation of both humans and hyenas to construct the significance of that place for both species. After the poisoning and subsequent showdown between the two clans, I referred to the place in front of Argobberi as an example of the "multispecies commons."[7] Whereas humans and other species are so often entangled in many subtle ways, the place outside the gate at Argobberi is like a sledgehammer. It's the kind of place where human and nonhuman socialities are historically, politically, culturally, ecologically, and religiously entangled, and explicitly so. If you want to see how humans and nonhumans mutually shape each other's worlds, then look at a place like the gate at Argobberi.

For me, the face-off at Argobberi was more than a revelation. It took me a step further toward membership in the Sofi clan. After Koti and the other hyenas arrived back at the feeding place, everything settled down. Some of the clan headed off to the garbage dump in search of food, while others lay down on the hill or beside the road. I went and sat on the stone wall opposite the hyena hole and waited. I wondered what might happen if a Sofi hyena went to the main market that night and encountered an Aboker hyena. After considerable waiting, during which I dozed off a couple of times, I saw a hyena cross the road from the forest and enter the drain that punctuated the wall just twenty meters down from the hyena hole. It was Dibbey. I got up and ran through the hyena hole nearest me. On entering Jugol, I made a left and hurried to meet up with Dibbey where that drain intersected with the lane. I found her waiting in the secluded darkness of the drain. She watched me approach and emerged when I stopped and waited for her, whereupon she crossed the lane and entered an open compound. The wall around this compound had largely disintegrated—the stones removed to build walls elsewhere—so it afforded easy entry and exit. Dibbey had found food here before, but on this occasion there was nothing, so she exited and joined me in the lane but was hesitant to go further into Jugol. I turned, offered an invitation, and headed in the direction of the drainage lanes that led to the main market. Dibbey followed, and all of a sudden I had a feeling of camaraderie with the spotted firebrand. About halfway up the lane I heard human voices and stopped in my tracks. Dibbey stopped alongside me. We both stared ahead at the intersection, illuminated by a yellow streetlight, and listened to the approaching voices. Two people appeared, but they turned right and headed toward Suqutatberi. The tension broken, we continued on toward the drainage lanes.

On arriving at the intersection, I saw a hyena approaching and went to see who it was. The hyena turned and headed down toward the hyena hole, so I turned around to rejoin Dibbey, but the stocky female had disappeared. I felt no loss, no disappointment; hyenas are pragmatic like that. I simply went on toward the main market, making the rounds of the lanes, avoiding people, and looking for other hyenas. Once I'd arrived at the main market, though, I felt a strong compulsion to seek human company. I called Nureddin, knowing he'd be awake from chewing khat all evening, and went to visit. His house was just around the corner from the main market, so before long I was there, sitting in my usual place in the living room, discussing hyenas, Harar's history, and my impending marriage to Tigist. Basically rehumanizing.

# 12

## A HOST OF OTHER IDEAS

When the rainy season set in, bringing mist and lush green to Harar, I married Tigist. For five months she'd rejected my invitations to dinner or burcha, and had only agreed to a date when my Harari friend Aman interceded on my behalf. Tigist was different from anyone I'd known anywhere. She'd grown up in the remote high country of West Shewa, where honesty, integrity, and family loyalty formed the very basis of a person's character. But she also had the will and determination of a moving continent. When she was sixteen years old and her father agreed to another family's proposal that she marry their son, she chose to resist. Before the sun had a chance to rise on the morning of her compulsory marriage, Tigist was up and running. The rest of the family was asleep after a night of wedding preparations, and Tigist made the most of her head start. She hopped from rock to rock to get across the river and ran to the road, where she hailed a bus. From there she traveled to Addis and then another five hundred kilometers to Harar, where her aunt took her in and kept the secret from her family. In Harar she faced an uncertain future, but it was one she could at least choose for herself, and one that, eventually, thankfully included me.

The events of our wedding were odd, to say the least. Abbas had proved himself a capable photographer with my still camera, so I gave him my camcorder to film the wedding. He inadvertently switched the device to night-shot mode; as a result, every scene in the first half of the video bore the same eerie green glow as my hyena footage. My wedding suit was a curious affair. I'd gone to a tailor in Andenya Menged with a picture of Gerry and the Pacemakers and asked him to replicate the 1960s-style suit, complete with stovepipe trousers. The end result was a suit in the standard Harari fashion, with baggy trousers that the tailor simply cut shorter. Meanwhile, the sleeves of the jacket extended almost to my fingertips. When I pointed this out to the tailor, he indicated that I should bend my arms. Problem solved. Our wedding car was a loan from the municipality. I was grateful for the loan, so I didn't argue when the driver refused to drive up Mount Hakim, where we planned to make our wedding vows. We ended up instead in a privately run garden. The curator there gave us fifteen minutes and demanded extra money if we shot any video or if the ceremony went beyond the fifteen-minute time limit. So we exchanged vows hurriedly, while members of the wedding party argued with the curator over the fee in the background.

By this time I was very close to Yusuf and his family. Yusuf took on the role of my father during the wedding, and Naime told her neighborhood group that her "son" was getting married. This obliged them to participate in the wedding preparations and celebrations, which they did with considerable enthusiasm.[1] Naime was delighted that I was marrying an Oromo woman, albeit a Christian one. After a formal reception at the Harar Cultural Center, including a cake mounted with fireworks, we had a celebration at Yusuf's house. Sixty people danced and sang in a hot, crowded living room, and crowds more spilled out into the compound. Mohammed, the Harari man whose wedding I'd attended six months earlier, had the idea that we should buy a load of meat and give it to the hyenas on the night of the celebration. The idea was ineluctably quelled by the afternoon's burcha, and the hyenas had to settle for their usual fare. Tigist and I left the reception in a friend's Bajaj, a fitting end to a quirky celebration.

Willi was quick to realize the change in my relational sphere. After our honeymoon, I brought Tigist with me every night to the feeding place. Over the first few nights, we sat together on the step of the shrine and watched the hyenas arriving. Willi, as always, came over and sniffed my knee and my boots, presumably to discern where I'd been during the day. On the fourth night, he finished sniffing me and began sniffing Tigist's knee. To her credit, she held her nerve and allowed him. Normally, when Willi approached people, he stood a couple of meters away and head-bobbed in front of them, trying to catch a whiff, but with Tigist he marched right up.

By this stage Willi and I were very close, but married life began to draw me away from him and the other hyenas. I began spending most of my time at the feeding place inside Yusuf's house with the family. Only after the feeding was over did I go out and sit with the hyenas on the hill. Then, by about 9:30 P.M., I went home with Tigist. The time I spent following hyenas in Jugol also diminished to almost nothing. After I was married, I stayed out late with the hyenas on only a couple of occasions. It felt odd to leave my wife alone at home while I traipsed up and down the drainage lanes in the company of hyenas. I also had to consider my personal safety, something I hadn't done too seriously up until that point. I had a responsibility to my wife not to get stabbed, shot, clubbed, or hit on the head with a rock while I was out at night. As a result, I felt myself drifting back into the human world, sleeping more during the night and less during the day, hosting visitors in the afternoons, and generally becoming more humanly social.

This change was compounded by people celebrating Ramadan. In Harar, the locals love Ramadan. It is a tremendously social time when many people sleep through the morning and emerge at dusk to go socializing until the wee small hours. Even at midnight the lanes and streets are crowded with people going about the town, carrying food, drink, and bags of khat. For the khat sellers, this means setting up shop and trading until well after midnight. For the hyenas, it drastically limits the time that they can spend in Jugol. During Ramadan, the Sofi hyenas remained at the feeding place until two in the morning, watching the crowds of people on the road between the feeding place and the town wall. Nureddin called Ramadan "waraba time" because of the way that people's active hours coincided more with the hyenas'. People became more hyenalike. As for the hyenas, they were frustrated in their attempts to make undisturbed forays into Jugol. And as for me, I found few opportunities to observe them doing so.

I wanted to keep my productivity up, so I organized a survey of people in Jugol and their attitudes toward hyenas. I made up a questionnaire and had it translated into Amharic, the official language of Ethiopia, which almost all people in Harar understood. The questions pertained to basic demographics such as age, ethnicity, and religion and finished off with a few simple questions about whether it was a good thing to have hyenas in Jugol and how the respondent felt about them. Arif, my Somali polyglot translator friend, had the task of collecting the responses. He selected a different subdistrict each day and went knocking on doors and intercepting pedestrians. He had an easy time getting responses and came back with some interesting results. I expected that Hararis in particular would have favorable views of hyenas, and that Muslims in general would appreciate them because they controlled the number of jinn. What I found, though, was that there was no discernible difference between the attitudes of Hararis, Oromo, Amhara, Muslims, Christians, men, women, or whomever, with regard to positive attitudes toward hyenas. Regardless of ethnicity, age, sex, or religion, the majority of people who responded were favorable toward Harar's hyenas. In fact, the only significant differences I found were between Jugol's subdistricts, known in Ethiopia as *kebeles*. Across all demographics, the people in Kebele 02 were the most appreciative of hyenas, while those in Kebele 05 were the least. I couldn't come up with any reason for this geographical variance. The only connection I could make between Kebele 05 and hyenas was that Kebele 05 was the quarter of Bedroberi (in Amharic, Buda Ber), where theriomorphic were-hyenas purportedly resided.[2]

At around this time I was contacted by Steph Dloniak, a researcher involved with the Maasai Mara hyena research project. She'd heard about my research and wanted to visit Harar to see the hyenas. I was in awe of the Mara hyena project; it was a behemoth. Beginning in the late 1980s with the work of Lawrence Frank, it continued under the direction of Kay Holekamp of Michigan State University. The project had NSF funding, a fleet of vehicles, radio collars, tracking devices, and a lab back in Michigan. Scores of doctoral students and undergrads had been fed through the workings, and they'd produced enough dissertations and publications to fill a hyena encyclopedia. From the outset of my research I'd been in contact with Kay Holekamp, who was by all accounts highly intelligent, very driven, and quite particular about the way the Mara project was run. She was also very encouraging and interested in what I was doing in Harar. It was she who had recommended me to Steph. But in comparison to the goal-directed research questions and rigorous data gathering of the Mara project, my wanderings around Jugol in the company of hyenas seemed amateurish. I invited Steph to visit and offered to put her up in my apartment, but in the back of my mind I couldn't shake the niggling feeling that my empirical emperor had no clothes.

We picked Steph up from the Ras Hotel in Dire Dawa. She had a serious disposition; her long hair, tied in a ponytail, and her boyish good looks gave her an air of professionalism. This impression was accentuated in no small part by the heavy bag of hyena-watching gear she'd brought along. We spent the drive to Harar talking about hyenas, hyena research, and conservation. "You know the Talek clan in the Mara has had only two alpha females since 1988." Wow, life histories of entire clans! Meanwhile, I couldn't even say which cubs of the Sofi clan were related to which adult. I tried my best not to sound too ignorant and focused on small talk. Steph lived in Kenya with her two small children and her Australian husband, who operated a safari company. She'd only had to endure a short flight from Nairobi and a connection in Addis, so she was neither tired nor challenged by the sights of Harar. I was expecting her to experience at least a little of the culture shock that so took me aback on my arrival, but she barely blinked. She knew what to expect of an African town.

Once she'd unloaded her gear at our apartment, Steph was eager to familiarize herself with the hyena landscape. I took her to see the dense vegetation surrounding the Sofi dens and across to the rocky outcrop where the Aboker hyenas raised their cubs. This answered a question foremost in her mind: how did hyenas raise cubs and find refuge in an urban environment? The Sofi and Aboker dens were relatively secluded places that were conveniently close to town. There was also a stream that cut through the suburbs of the New Town, along which the three-legged hyena, Bajaj, apparently had her den. For Steph, this was very unusual. Harar's hyenas were living in places where houses, fences, and the like were right on top of them. But that was just the tip of the strangeness that she was set to see in Harar.

The setting sun took us to the Sofi feeding place. We'd already visited in the afternoon. Steph had brought candy for Naime, presents for Yusuf's kids, and photos of the Talek hyenas and their cubs to show Yusuf. In the Maasai Mara, the researchers could drive right up to the dens and almost point their cameras down the holes to photograph cubs. The hyenas there were apparently accustomed to having vehicles full of researchers circling around, following them on their hunts, and locating them in the damnedest places via the radio collars. Take a person out of the car, though, and the hyenas ran for the hills. They made a clear distinction between a bunch of human heads peering at them through car windows and complete human figures standing on the ground. The former were familiar and relatively innocuous; the latter, terrifying. The one exception was a Maasai tribesperson sitting in a vehicle. This was apparently as scary to hyenas as people on foot were. Steph was critical of the Maasai for their chauvinistic oppression of women, their fanatical accumulation of cattle, regardless of means or rainfall, and above all for their murderous hatred of hyenas. She was not a little bemused to see the Harar hyenas trotting up to take food from a benevolent hyena man sitting on a rock, flanked by tourists.

When the feeding was over, I invited Steph to go and have a look at the hyenas on the hill. She'd brought along a flashlight as big as a baseball bat and a little brighter than a landing light. It lit up the hill like a stadium. My flashlight envy was fleeting, though, because the Sofi hyenas all scattered when Steph followed the beam of her Maglite Maximus onto the hill. This was interesting. Apparently, the hyenas in the Mara were unfazed by the beam of a flashlight or a set of car headlights and went about their business undisturbed. But the Sofi hyenas had different attitudes toward lights. In circumstances other than feeding, flashlights made Sofi hyenas nervous. Even the dim LED flashlights that the locals used to navigate Jugol at night made them uncomfortable, because they knew there was a human on the other end. The Sofi hyenas taught me early on to hold my flashlight at the far end so that I could cover the lens with my hand and let out just the tiniest amount of light. And even then I had to point it at the ground near, rather than directly at, the hyenas so as not to disturb them.

The following morning we woke at 4:00 A.M. to go looking for hyenas in Jugol. I was particularly concerned about Steph's safety, so I made sure we stuck to the metaled roads under the streetlights. But Harar confounds the soundest of intentions, and the power soon went out. In the light of our flashlights, we found Willi and Koti paying attention to a man outside his gate on Suqutatberi road. The man was calling to them and pretending to have some food, and Koti and Willi had fallen for his ruse. Sometimes people did this to hyenas for whatever reason, but sometimes they actually had food. The hyenas had no choice but to go along with it until they either got the food or decided there was none to be had. Eventually, Willi and Koti gave up on the man and headed downhill toward the Suqutatberi gate. We followed, but a cold wind quickly kicked up, followed a minute later by the inevitable downpour. Willi and Koti disappeared into the farmland, while Steph and I stood under an awning, waiting for the rain to subside. A homeless man huddling in a niche pulled his tattered blanket further over his head as the rain soaked him through.

After the obligatory few minutes of torrential downpour, the rain eased up. We followed Harar's outpouring of water to the outer road and made our way around to Argobberi and Aboker territory. I'd figured on the playing field as the best bet for seeing hyenas in the early morning, and it turned out I was on the mark. There must have been half a dozen hyenas of the Aboker clan competing with the dogs and cats for scraps from Sisay the butcher.[3] Another six hyenas were obviously not hungry but content to hang around. The three-legged hyena, Bajaj, was social-sniffing with two others, while another two were lying down on the bank beside the stream. One skinny male was running between two Amhara boys who kicked their ball back and forth. He was fascinated by the game, but not so game as to get directly involved.

The sky was overcast, but the gray light of the morning had spread across the field. Surrounded by hyenas, Steph was in her element. She voiced her observations as though speaking into a voice recorder. Here, appeasement by the subadult; there, vigilance from Bajaj because of the dog; over there, displacement and, again, appeasement. She quickly worked out the ranking relationships among the Aboker hyenas present and the dominance of the one I called Big Red and her friendship with the hyena I called Flopsy. These two were very close, very often together. Steph had all the observational skills and terminology to make quick sense of all the goings-on at the field, but the hyenas refused to allow her to remain a distanced, objective observer. I noticed a wondering smile break across her face when a young hyena approached and stood before her, silently asking what she had to offer.

Back at the feeding place that night, I took Steph out of her element. Yusuf had finished feeding the hyenas, and we all sat at the shrine watching the last of them scouring the ground for scraps that had gone unnoticed in the excitement.

Over between the road that ran through the feeding area and the drainage canal, I saw the silhouette of a hyena lying on the grass. I assumed that it was Willi because that was his favorite place to lie down. By the time I'd walked halfway across the feeding place he still hadn't budged, so I knew for certain it was Willi. By that stage of my research, I always carried a plastic comb in my jacket pocket. I needed something long and chewable to distract Willi from biting my arm, and in lieu of any kind of dog toys in Harar, combs sufficed nicely. He loved chewing on them and demolished one every few days. But then it occurred to me that I could use the comb for its intended purpose, and I began combing Willi's fur. When I first did this in front of Baby, her reaction was priceless. She lifted her head and gawked at us like a shocked headmistress. Willi loved being combed, though, and he yawned uncontrollably when I combed under his chin.

I sat on the ground beside Willi and took the comb from my pocket. He turned his head and snapped at the comb but after I pulled it away several times he allowed me to comb his forehead. Steph was as incredulous as Baby had at first been. In the Maasai Mara, if hyenas weren't running away from people, they were eating them. Under no circumstances did they permit people to march up and comb the fur between their ears. I called Steph over, not thinking for a minute that the approach of a stranger might frighten Willi away. She stood and walked slowly over, while Willi and I remained on the ground. I handed the comb to Steph and she gave Willi a little tickle under his chin.

Willi had a talent for making people uncomfortable. Often, when a tourist was standing beside Yusuf, holding the feeding stick for the hyenas, Willi went around behind them and sniffed their rear ends. Needless to say, this made people pretty nervous. One night Willi walked up to an arriving Bajaj and sniffed at the front wheel. The tourists inside were so frightened that they refused to get out and join in the feeding. Instead, they watched from the backseat, held captive in their own vehicle by a young hyena. He also tended to approach people walking around Jugol. People didn't like that, and they stamped their feet to chase him away. But Willi took Steph out of her comfort zone in an entirely different way. Here was a singular hyena adopting a singular trajectory.[4] He was an outlier, doing strange and significant things that would otherwise fall off a regression line. In the shadows of the feeding place, Willi was a someone who was difficult to reduce to predictable, species-typical behaviors. Steph returned the comb to me. We stood up simultaneously—which made Willi a little nervous—and returned to the step of the shrine.

Steph had one more day in Harar, so in keeping with my five-star treatment of guests, I took her to see the garbage dump. When we arrived and started photographing the hyenas, four young Oromo guys ran up and stood between us and our subjects. They threatened us and demanded money, saying the hyenas belonged to them. I invoked the name of the most intimidating person in Harar—the director of tourism, sport, and culture—and the four ran off, leaving us to continue in peace. The garbage truck came and went occasionally, each time depositing a load of fresh garbage. An indigent waited for each new load and began sorting through the garbage before the truck had even completely tipped the dumpster. The indigent was mindful of the hyenas. He withdrew bones and bits of skin that he unearthed and threw them over his shoulder to the waiting hyenas. Steph was noting the goings-on: the hyenas being especially vigilant when people disappeared behind mounds of rubbish, this one submissive to that one, these two squabbling over the same bone. At one point Baby emerged with a sheepskin, another hyena hot on her heels. Baby giggled and raced down the dirt track with her head high, ears pinned back, and the skin trailing between her legs. Steph commented on how this behavior was more like that of wild hyenas. Harar's hyenas were indeed wild, but I knew what she meant. The wild hyenas she was referring to were the kind that biologists, tourists, and natural history filmmakers preferred, the ones that chased after wildebeest and competed with lions over zebra carcasses. Those were the real hyenas. The ones in Harar were shadows of the ecologically authentic hyenas that one expected to see in Africa.

After fewer than twenty minutes at the garbage dump, Steph indicated that she'd seen enough and wanted to get out of there. This took me by surprise, because I didn't see the garbage dump as a disturbing place. Even the smell was not unpleasant; it was kind of earthy. I saw the garbage dump as a nice, secure place that was full of opportunity, an ideal spot for spending time with the hyenas. The atmosphere was relaxed, especially in comparison to the feeding place or the lanes in Jugol, and there was food. Normally, I would have stayed with the hyenas until they decided to go home, but instead we packed up and left. Whereas I'd expected that the living conditions, khat addicts, and misshapen humans would most challenge my biologist friend, it turned out that her greatest culture shock came at a place where hyenas successfully eked out an existence among mounds of urban waste. Admittedly, though, on that particular morning the smell was a little less earthy than it was garbagey.

After Steph returned to Kenya, only a few days remained of my time in Harar. I spent my mornings at the playing field and the garbage dump, being with hyenas as often as I could. Willi had originally been nervous with me when we encountered each other in daylight. Hyenas are quite aware of the advantage they hold over humans in darkness, but in daylight they become very wary of people. Nevertheless, Willi got used to the daytime me, and we hung around together at the garbage dump until after 7:00 A.M., when the sun was well risen. By that time, the imperative to return to the safety of his den took over and he left for home. Meanwhile, I wandered back to my own home for breakfast and some sleep.

My time in Harar was up. I'd extended my plane ticket to the edge of its tolerance, and a conference the following month in Tasmania had my name on its list of presenters. I was due to give a talk about hyena calls and those people in Harar who understood hyena language. Our plan was for me to fly back to Australia and for Tigist to follow shortly afterward on a visitor visa. We could wait on the outcome of our marriage visa application after that.

The night before I was due to leave, we went together to the feeding place. Unlike Harari houses, Yusuf's house had only a single-level platform that encircled the room, and a rammed-earth floor inside the door. I sat with my wife and my adopted family, handing out presents and posing for photos. Naime brought in a spicy-hot chicken curry, so hot it was remarkable that the meat had remained intact among the lavalike spices.[5] Chickens in Harar are far tougher than their Westernized cousins; they grow more slowly, range freely, and have no need for human intervention to defend them from cats and bacteria. The food was far too spicy for my mollycoddled taste buds, so I had a single mouthful and went outside to say good-bye to the hyenas.

Abbas had been delegated the feeding duties for the night, so he was happy when I brought a little of the celebration outside. We sat on the step of the shrine and chatted in our concoction of Harari, Oromo, and English. Abbas asked when I'd be returning, what would happen when I got to Australia, how many children we'd have—all kinds of questions to which my stock answer was *an ukhumekh*. I don't know. Some tourists arrived, Abbas stood up and carried the food bucket to the center of the feeding area, and Willi trotted around from behind the compound.

The hyenas must have had their fill of some other food the previous night, because they had little interest in feeding. Kamareya lay down with his chin on a stone and watched the feeding with an indifference that only comes with a full stomach. The others wandered around the feeding place as though they were there out of habit rather than hunger. The only one interested was Willi. He stood with his chest against the feed bucket and took every single scrap doled out by Abbas and the tourists. That is, until his attention span reached its limit and he ambled off to the hill, with Abbas calling after him. I got up and followed, hoping to say my good-byes.

I arrived at the hill in darkness and scanned the scene with my night-vision scope. A couple of young hyenas got up and moved further up the hill, disconcerted at having to share the space with me. The others remained where they were while I tried to determine which was Willi. As it turned out, Willi had gone. I supposed he'd gone off to the garbage dump, or else out into the farmland or back to the Sofi dens. The options were too many to justify my going to look for him, so I sat down on the grass instead and contented myself with being among the others.

Once seated, I saw the silhouette of a hyena ambling toward me. As the hyena came closer, I made out in the darkness the pointy face and little beady eyes of Baby. Over the months, she'd changed in her attitude toward me. She'd gone from indifference, to active avoidance, to engaging me in play. Now she'd seen me sit down on the hill and had deliberately come over to lie down beside me. She arrived, turned around to face the same direction as me, and lay down at arm's length.

I had Willi's comb in my pocket. I'd intended to give it over to Willi to destroy, but I figured Baby would appreciate the gift. She loved the manufactured artifacts of the urban environment. I took out the comb, leaned across, and held it in front of Baby's nose. Immediately interested, she sniffed at the comb and followed it to the ground when I put it down. She put a paw on it and began chewing. Judging by her enthusiasm, I anticipated that by morning there would be little more than a scattering of red plastic among the fragments of broken glass on the hill.[6] I got up, said good-bye to Baby, and left the hyenas.

The next day I went to Dire Dawa airport with Tigist and Yusuf in a minibus the mayor had provided. It was ironic; I was leaving Harar in a relatively empty minibus but with another kind of fullness. I had a wife, an adopted family, and membership in a hyena clan. I expected I'd miss them all. We spent the obligatory several hours waiting for security checks at Dire Dawa airport, and by midafternoon I was airborne over the Rift and headed for home.[7] I was excited about the future but felt weighed down. My head and my heart were both telling me that I'd never see Willi and the Sofi hyenas again.

# 13

## RETURNING TO OTHER HYENAS

My doubts about ever returning to Harar were quickly overturned by the Australian immigration department, which refused Tigist a visitor visa. After only three months in Australia, I flew back to Ethiopia to be with Tigist while we waited on the outcome of our application for a marriage visa. We paid a quick visit to Tigist's family in West Shewa and then took a bus across the Rift back to Harar. While I was away, Tigist had been staying with a friend of ours named Arianna, an Italian NGO worker who had a house in Jugol. Arianna was tall, blonde, and stylish, with designer eyewear and a purebred retriever. These kinds of attributes are a poor fit for Harar. Frustrated by the apathy and intransigence of the Hararis, Arianna had given up hope of effecting development in the town and was spending most of her time with her husband and adopted children in Addis, waiting impatiently for her contract to expire. We had the house in Jugol entirely to ourselves, apart from a couple of occasions when Arianna came to visit. She said that she envied my working with hyenas. Not only was she frustrated and disillusioned by government officials and bureaucrats; she was exhausted by the stone-throwing children, the hucksters, the khat addicts, and all manner of challenging humans.[1]

I was happy to be back in Harar and excited about getting back to the feeding place and the hyenas. I broke into a run when I arrived at the bottom of the hill below Argobberi and had a warm reunion with my adoptive family. Ajebbo and Nasruti clung to me like limpets, while Ardalle, as befitted a teenager, restrained herself and hugged Tigist instead. Naime went all teary-eyed and Yusuf beamed like a cat. Over the months I'd been away, he'd given up khat and cigarettes, so his serious disposition had vanished, replaced with a jovial, sociable character. The kids were savoring their khat-free dad like the sun breaking through the clouds.

We shared a meal and spent the evening in the house, talking about the past three months and handing out presents. Every now and then a tour guide appeared in the doorway asking that either Yusuf or Abbas come and feed the hyenas. Recalcitrant as ever, Abbas made a show of his reluctance to go out and feed the hyenas in the middle of a family reunion. He returned after only a few minutes, and I'm sure that the tour guides were disgruntled at that and the tourists disappointed. I wondered if the hyenas were getting frustrated as well.

I was itching to go out and have a reunion with the hyenas, but it would have been poor form to abandon the family too early. I waited until after 9:00 P.M. before putting on my boots and going out into the night to see who was still around. I knew that Willi was still coming to feed—Abbas had told me as much—so I went directly across to Willi's favorite place beside the drainage canal in the hope of finding him. As I got closer, I made out the silhouette of a hyena lying down with his chin on his carpals. The hyena didn't even turn his head when I was within three meters, so I knew for certain it was Willi. I went right up and sat beside him on the grass. His ears swiveled backward to follow my movements, but he didn't turn to look at me. I spoke quietly and noticed that he cocked an ear to listen but still he remained with his chin on his carpals, staring straight ahead. We both sat quietly for a while before the sound of some hyenas squabbling in the forest behind us broke the silence. Willi turned his head to face the noise and took the opportunity to look at me for the first time. It wasn't a quizzical or curious look; it was a stare. With his nose down, he simply fixed me with his eyes before turning his head and returning his chin to its former position. Some reunion. I stood up and wandered over to the shrine feeling ignored and rejected.

In 1989, Kay Holekamp and her colleagues at the Maasai Mara hyena research project noted with interest the departure of fifteen hyenas from the Talek clan.[2] Eight adult females of three matrilines and their dependent offspring abandoned the Talek home range. They set up shop on the far side of the Talek River, in an adjacent territory that had become vacant owing to large-scale poisoning by Maasai tribesmen. But there was also some food scarcity in their own home range, and these mid- to low-ranked hyenas were always going to be near the back of the feeding queue whenever a Talek hyena made a kill. The researchers named this splinter group the embryo clan and noted the completeness of the separation when the embryo hyenas one night engaged in a full-blown clan war with their former clan mates. It seemed that they had broken away for good, but not so. After about eight months, three females from one of the breakaway matrilines returned to the Talek clan and paid a high price for their disloyalty. These girls plummeted to the bottom of the Talek hierarchy and suffered attacks by the resident hyenas at levels of aggression previously unseen under any circumstances. Either the Talek hyenas considered these returnees unwanted immigrants or they considered them returned traitors.[3]

It occurred to me that Willi was miffed that I'd gone away for three months; that would certainly account for my cool reception. Theoretically, he might even have chosen to attack me in the fashion of the Talek hyenas, but I'm not so sure about that. In the case of the returnees from the breakaway embryo clan, Talek females attacked returning females, while the Talek males seem to have taken no part in the retribution. Willi's response to my return was less than aggressive but more than indifferent. He turned to listen to the hyena sounds from the forest and chose that moment to look me in the eye. It was a cold, hard stare, and it left me feeling chilled; it was enough to persuade me to get up and leave.

The following night, I returned to the feeding place wondering where I stood in relation to Willi. As dusk set in, Tigist went into the house with the girls while I went to visit Naime, who was in the cooking hut. I found her in front of the cooking fire, in a haze of smoke that poured from the hut. She was busy making *ukhat*, a soft, spongy kind of fermented pancake that tastes sour on its own but delicious when it's allowed to absorb the flavor of the curry that's poured on top.[4] I sat on my heels outside the hut and talked through the open doorway as Naime mixed, poured, and covered each ukhat before removing it from the hotplate and placing it on a growing pile to cool. As it grew darker, I noticed a hyena sniffing around at the bottom corner of the compound. It was Willi. He sniffed around the latrine, then crossed between the buildings and looked into the living room hut where Tigist and the girls were. He knew that I was in the middle of the compound, but not once did he look at me. I turned to watch as he passed me and went to the other end of the hut, where he peered through the gate at the feeding place. Then he turned, came straight over to where I was squatting, and stood right beside me. Naime exclaimed in Harari, "Allah! Is that Willi?" I said yes and began scratching Willi's rump. He turned, sniffed at my sleeve, took a few bites, and then led me over to the step of the living room hut, where he picked up a PVC sandal. I said, "Hey!" and he knew something was up because he ran off with the sandal in his mouth and me in hot pursuit. I ran after him across the compound to where he found a hole in the thorn fence. He disappeared, and I had to run all the way around to the hill, where I found him chewing on the sandal. We were reunited.

Hyenas are the Yorkshiremen of the carnivora. They're a dour lot and not given to the displays of affection we're used to seeing in cats; they don't rub up against one another, squinting and purring; they don't meticulously lick each other's coats. That's not to say that hyenas have no conception of bonding or affection; it's just that their conception of these things is different from that of humans and cats. When Willi came and stood beside me, it was as meaningful and touching for me as Ajebbo's hugs, Naime's tears, and Yusuf's smiling embrace. To stand beside another, to face in the same direction as another, for a hyena is an acceptance, a declaration of solidarity, and, in the context of the previous night's snubbing, an expression of forgiveness. And Willi did all that without even making eye contact. This is hyena. They commune in the most subtle of ways, using an embodied language the vocabulary of which extends beyond a hyena's body and into the space and temporality of the world. When you're a hyena, it's not just about what you do with your body; it's also about where you do it and when.

As night followed night at the feeding place, I couldn't help noticing a gap in the suite of hyena personalities. Baby was missing. At first I hoped she was simply stuffed from a windfall, maybe a dead ox found out in the farmland. I thought she might come back when her hunger returned, but after four nights I knew there must have been more to her absence. Sitting at the shrine beside Abbas and Yusuf, I asked Abbas if Baby had been coming to feed lately. Without looking at me, Abbas simply waved his index finger from side to side to indicate a negative. We sat in silence for a while before Abbas asked Yusuf, "Dad, what do you think happened to Baby?" Yusuf responded, "I don't know. Maybe the other hyenas chased her away." Dibbey happened to be lying in front of us, staring up at me with her beautiful, deadly-brown eyes. Abbas asked, "Do you think they'll chase Willi away?" Yusuf paused for a moment. Normally he would have puffed on a cigarette before answering, "They can do it with any hyena."

That was it. No more Baby. If she had been a he, I would have suspected that Baby had emigrated according to the biological imperative of mature hyena males.[5] Was I mistaken in determining Baby's sex? After all, the morphology of hyena genitalia makes it hard to tell females from males. But I had a clear photo of Baby's dangly phallus from when she was social-sniffing with Willi and Kamareya, and it looked unequivocally like a female's.[6] Given that females usually stay within their natal clans, I could only conclude that Baby must have died. In Harar, this was a more likely scenario than emigration; there were snares, poisons, busy roads, killer farmers, and all manner of dangers in the landscape around the town. And while a careless or overly bold hyena might do well to get food in a landscape full of humans, that hyena would also be quite vulnerable to the many means that humans have of inflicting death on wildlife.

Since my return, I'd noticed some other changes among the Sofi clan. Willi was noticeably larger. I expected this, but he really had put on weight. I guess it paid to be at the front of the queue at feeding time. There were also a few young newcomers at the feeding place, including one very nervous juvenile who was too afraid to feed and hung around on the periphery. I was intrigued by this one because she often accompanied Dibbey. Once, when Dibbey ran to get a bit of meat that Yusuf had thrown out into the darkness, the little hyena tried to nudge Dibbey out of the way and get to the food. This she did without consequence. The next time she tried this, Dibbey actually moved aside to allow the little hyena access to another piece of meat. In lieu of a name from Yusuf, I called this one Muka Dibbey, meaning drumstick. Even if she wasn't Dibbey's cub, she certainly knew how to beat the drum. There was another youngster I'd seen on the hill before I returned to Australia. This one was very bold indeed and stood very close to Abbas when he was feeding. He called her Fintamurey, meaning offcut. From what she was cut, I knew not, but it must have been audacious stuff. Fintamurey was one very bold and, it turned out, unsettling hyena.

My first up-close encounter with Fintamurey came after about two and a half weeks back in Harar. I was sitting on the hill in the darkness while the majority of the hyenas lay scattered about. Willi had trotted off down the road toward the farmland and left me sitting by myself. I sensed a hyena behind me and turned to find Fintamurey head-bobbing and sniffing the air. She tried to get around back of me so I pivoted on my heels as she circled around. I took out Willi's comb and offered it up. She took a few bites and eventually let me scratch her chin with it. Still, she tried to get around back of me, so I got up and went to the fence of Yusuf's compound to protect my back. Once I was there, she gave up on the comb and tried biting me. She bit at my arms and legs, and it was all I could do to fend her off. Fed up with the biting, I stood and went into the living room hut, where Naime and the girls were watching TV. I sat inside on the platform and noticed Fintamurey at the doorway staring in at me. Nasruti's efforts to shoo her off were in vain, so I went out and led Fintamurey over to the back of the shrine. There, in the light of the street lamp, I hoped she might be a little less bold. No such luck. She took a bite of my sleeve and then, bizarrely, began licking it. Another bite and more licking and I gave up on my hyena watching for the night and went into the house. Fintamurey was just too pushy.

Within weeks, Fintamurey was verging on the obnoxious, and she was leading Willi astray. While most of the hyenas stood at a distance from Yusuf or Abbas when they were feeding, and the others stood just in front, Fintamurey, and then Willi, began sticking their heads in the feed bucket. This was annoying for Yusuf, who had to put his hand in the bucket to extract bits of meat for the other hyenas. It was pretty crowded in there, with a couple of hyena snouts and their bone-crushing teeth blindly chomping on the bucket's contents, and Yusuf valued all five of his fingers. He had to constantly push the two pushy hyenas away, saying, *Achi dem, achi dem* (Get away, get away!). Nor was I immune to the audacious duo. The next night, when Willi came to sniff me where I sat at the shrine, Fintamurey joined him and put her snout between my legs, hyena fashion. I might have trusted Willi to do this, but Fintamurey was a loose cannon. A set of sharp teeth on the end of an unpredictable head sniffing my crotch was too disturbing to bear, and I had to lift Fintamurey by the neck and shove her away. Then Willi joined in, and I had to shove them both away several times before they finally gave up.

I felt that I'd created a monster. Fintamurey as a young cub had watched as Willi and I played, combed, bit, and sat together on the hill. Little wonder, then, that she had decided I was connected to the Sofi clan and wanted to introduce herself. Hyena cubs do this with older hyenas. In addition to introducing themselves to their elders, they must learn one by one the individual scent and rank of each member of the clan, and they do this with a good sniff of the genitals.[7] It was little consolation knowing that she was unlikely to bite mine, but it was very disturbing to see that she had almost no fear of humans and to imagine that this was influenced by my relationship with Willi.

I had to put aside my misgivings about Fintamurey for a couple of weeks. Steph Dloniak had invited Tigist and me to visit her in Kenya and see the Maasai Mara hyena project. Not only was this a great opportunity to see a different bunch of hyenas doing different things from those in Harar; it was a chance for a honeymoon that didn't entail sleeping in a hut with an extended family. We flew out of Addis over the Rift Valley's blue lakes and parched ground, like a turquoise necklace left in the desert, and just over an hour later landed at Nairobi airport.

Anyone interested in seeing what happens when a transport network breaks down would be hard pressed to find a better example than Nairobi. The outskirts of town aren't so bad, but with proximity to the town center, the traffic becomes proportionally worse. Where vehicles usually sit backed up for kilometers from intersections, minibuses drive on the wrong side of the road and try to duck back into the lane when vehicles come from the opposite direction. This tends to block the traffic coming from the other way as well. As for roundabouts, these are always blocked, so giving way is not an option. Drivers simply drive right on in, trusting in the reluctance of other drivers to T-bone them. The bizarre and sad reality of it is the apparent absence of any alternative. Unless a driver does what everybody else is doing, not only will he make no progress, but he'll also make a lot of people behind him quite irate. We were lucky to have a local driver to see us through. Steph's husband, Howard, assigned one of the guides from his safari tour company, a Kikuyu man named William, to pick us up and deliver us to their house on the outskirts of

Nairobi. William navigated the chaos deftly and efficiently; by the time we opened our eyes, we were stopped in front of Steph's house, where a welded metal hyena watched us from among the plants in a garden bed.

After a night in Nairobi, we woke early to drive out to the Maasai Mara. Tigist and I left with William in one of the safari vehicles, while Steph was to fly out later in the afternoon and spare herself hours of back pain. Driving down the escarpment into the Rift Valley, I probed William on his thoughts about wildlife in Kenya, in particular hyenas. On the subject of human/wildlife conflict, he insisted that relocations were the solution and that the research he was aware of backed him up. Of hyenas, he said that he'd previously disliked them, as most Kenyans do, but that Steph's enthusiasm had rubbed off. He now found them fascinating.

We stopped for lunch in Narok, a town that prospered on the back of tourism, specifically, visitors to the Maasai Mara. The most obvious tourist ventures were the curio shops that sold carved animal figurines, although none sold hyenas. William explained that the wood-carvers imagined that nobody would ever pay for a carving of a hyena. Moreover, hyenas were associated with sorcery and Kenyan witchcraft. An effigy of a hyena could well bring misfortune to its maker. We were sitting in a pleasant café under the shade of some trees when William took the opportunity to explain the workings of the Maasai Mara reserve. Unlike Kenya's national parks, which are administered by the Kenya Wildlife Service, the Maasai Mara is a reserve divided into three units, administered locally. The west is administered by the Mara Conservancy, a Maasai nonprofit organization, the outer areas are controlled by Maasai Group Ranch Trusts, and the east—the part we were visiting—is administered by the Narok County Council. William explained that some vested interests were having negative effects in the eastern part of the reserve. I asked what kind of effects and William said merely, "You'll see." He also explained that there were controls on the number of tour vehicles allowed and that walking in the reserve was prohibited. People walking about the place would only disturb the animals or else be eaten. And that was bad for tourism.

The road from Narok to the reserve was like an eighty-kilometer-long bombing range. This didn't deter William, who gunned the engine as we careered down the road dodging large, abundant potholes. We spent 30 percent of the drive with two wheels on the dirt shoulder and two on the pockmarked road. The other 70 percent we spent with all four wheels on the dirt shoulder. The remnants of asphalt served simply as a divider between us and vehicles traveling out of the Mara on the opposite shoulder. Often, one of the shoulders was too rough, as rough as the asphalt, so vehicles traveling in both directions used the same shoulder and passed within inches of each other at heart-stopping speeds. I shouted above the screaming engine and thumping, rattling suspension, Why is the road so bad!

William shouted in reply, "It's the responsibility of the Narok County Council!" That explained nothing. I asked, So why don't they fix it! "Some of them have an interest in charter flight companies! If they fix the road, then nobody will have reason to fly into the Mara! They'll come by car from Nairobi!"

As we crossed into the reserve, the road turned to smooth, quiet dirt. It was like sailing out of a storm. A few wildebeest and zebras crossed in front of us, and William explained that these were residents. There were others who only visited during the annual migration, but they had gone south to the Serengeti. At the top of a rise, we encountered a couple of tour vehicles parked beside a thicket. A cheetah was dining on a Grant's gazelle in the shade of the thicket while tourists stood in their open-top tour vehicles taking photographs. The cheetah looked over-heated but otherwise unfazed by the attention. Large carnivores are used to being stared at. William explained that at peak season—during the migration—there would be at least a dozen vehicles surrounding this cheetah. The most he'd ever seen were eighty-two minibuses and four-wheel-drive vehicles gathered around a cheetah kill. And I'd thought hyenas were rubbernecks.

The hyena research station was located at the northern end of the reserve, just beside the Talek River. On arrival, we were greeted by Laura, a petite, pony-tailed undergrad whom Kay Holekamp had employed to look after the project in her absence. There was only one other research assistant there; that was Mike, an athletic-looking undergrad with close-cropped hair and a Ramones T-shirt. Another two Americans, Serena and Troy, friends of Laura's on vacation, were staying at the camp. They were less interested in the hyena research than they were in capturing photos of the abundant wildlife. Then there was the Kenyan camp cook and a man who took care of security. His primary task was to ensure that researchers and visitors weren't killed by hippos.

No permanent constructions were allowed in the Maasai Mara, so the re-search station amounted to a collection of tents set up for a long-term stay. We stayed in the tent belonging to Kay. Its interior was very homey, with a bed, a desk, and a chest of drawers, which together made the tent seem more or less perma-nent—not surprising given that the camp had been there for more than twenty years. We settled in and waited for evening. I was eager to encounter some Mara hyenas and see how professional hyena research was done.

Steph arrived in the afternoon. She'd come to accompany us but was also hoping for an opportunity to instruct Laura on darting technique. It would re-quire the right circumstances, though: a hyena alone and unaware of the connec-tion between the sudden pain in his rear end and the research vehicle fifty meters away. If another hyena happened to witness a darting, a research vehicle driving up alongside, and a couple of researchers getting out to weigh, sex, and collar her clan mate, she would be sure to avoid that vehicle in the future. I recalled Hans Kruuk's comment on hyenas' wariness of his research vehicle after he tranquilized them and cut identifying notches in their ears.

We set off an hour before sunset. The tick-tick of a receiver on the dash betrayed the location of one of the clan's collared hyenas, and Laura swerved the vehicle left and right, looking for the strongest signal. We found the collared hyena with three others in a clearing beside a thicket. Their tails were up like masts as they stood shoulder to shoulder, facing the trees and bushes. Laura called rapid-fire observations into a voice recorder as she flipped through the pages of the heavy reference book containing left- and right-profile photos of all the Talek hyenas. Until she made an identification, she simply called them hyena number one, hyena number two, and so forth. I asked what was happening. "There's two lions in there," said Steph. "These guys are pretty excited, but nothing much will come of it." She was right. After a while, the hyenas dispersed. One or two remained, and we stayed with them until darkness set in. The lions never emerged, but a glowing pair of eyes nearby betrayed a leopard, who descended from a tree and crossed the clearing, oblivious of the lions and hyenas. As we watched the leopard disappear into a thicket, Laura fired up the engine. Where to now? "Back to camp." It was only 8:30 P.M. and the hyenas had only just started their rounds. Don't you ever follow them at night? Steph fielded my question. "Joe Kolowski used to. He got out here with his night-vision equipment and followed them around in the dark. We don't do that so much."[8] In the Mara project, most of the research questions pertained to things like rank relations and ontogeny that could be answered by visits to the communal den and by tracking hyenas in the early morning and evening. The corollary of this was a good night's sleep for the researchers. I was envious.

We returned to the research station and enjoyed a dinner of egg and bean lasagna prepared by the camp cook and served up in the mess tent. Naturally, the conversation centered on hyenas. We talked about the findings of the Serengeti research project and the cases of adoption they'd reported among the hyenas there. Steph was doubtful; in the history of the Talek research project there had been zero adoptions.[9] When I mentioned to Mike my method of following hyenas in Harar, he was intrigued. He confided that he sometimes went jogging in the reserve in the morning but had never gone hyena watching on foot. It was something you just didn't do in the Mara.

The following evening, we set off again in the research vehicle, looking for Talek hyenas. Standard practice when looking for a hyena with a radio collar is to drive around until you pick up a signal. This you do with an omnidirectional receiver, which detects signals from anywhere in the vicinity. Once you have a mark, you change to a unidirectional receiver and drive in a circle. The signal should sound only when the receiver—and the vehicle it's mounted on—is pointing toward the collar. You find the signal and drive in that direction. Simple. This works well in the Maasai Mara because it's mostly grassy plain, with few creeks or woods to get in the way. In Harar, you'd be driving into walls.

As we drove around the plain, Tigist proved herself an astute hyena spotter. "Over there!" she'd say, and the rest of us would turn and stare, trying to see the hidden hyena. Only when we drove closer did we find the one she'd seen. In one case, we failed to spot a hyena until we were only a couple of meters away and she got up from her ditch and wandered off. I'd expected the hyenas to be habituated to the vehicle, but some obviously found us intrusive, like a carload of boorish guests at a hyena cocktail party.

We followed one signal to a collared hyena who was resting under a tree. The hyena looked at us and I could see her sniffing the air. She got up and walked off, so Laura stepped on the gas and followed. By this time it was getting dark, and I hoped to see some hunting.

Our mark dropped down into a dry creek bed, so we drove past and forded the creek farther on, hoping to meet the hyena when she emerged. It was quite a wait, but when she finally emerged she was in the company of five other hyenas. All six of them headed off purposefully into the darkness, with us trundling along behind at an idle. I recalled Hans Kruuk's description of hyenas teaming up in groups of six to hunt zebras in Ngorongoro Crater and wondered if our group was doing likewise. Sure enough, a mass of black-and-white stripes soon appeared in our headlights. There were eyes everywhere reflecting the light, and our hyenas were intent on unsettling the zebras. They cantered as a group toward the herd, and the zebras immediately ran off. The hyenas followed at a steady pace, crisscrossing behind their quarry, not gaining on the zebras but not losing ground. After a hundred meters of this peculiar chase, the hyenas slowed to a walk and let the zebras go. Laura drove circles trying to locate the collared hyena, but she'd disappeared in the confusion. We headed off into the darkness and circled around for half an hour to no avail. Again 8:30 came around, and we headed back to camp.

Steph was flying out the next day and asked if we wanted tickets. We were ready to leave. Tigist was bored with the place, while I felt disappointed and disconnected. Everything had to be experienced through a car window; the only place where our feet touched the ground was in camp. Even when Laura collected hyena poop, she drove up alongside it, leaned out of the car, and picked it up without setting foot on the ground. All in all, it felt like watching a particularly boring natural history documentary at a drive-in theater.

Morning came, and it was time to go for one last look at the hyenas. Steph was staying in camp, getting ready to leave, and Laura was taking her friends for a balloon flight with one of the tour operators. That left Mike, Tigist, and me. We had a quick breakfast, climbed into the white pickup with a hyena decal on each door, and headed out across the plains in the darkness of early morning.

According to Mike, the Talek hyenas had relocated their communal den and the researchers had yet to find the new one. This was presenting a problem because of the importance of the communal den to the hyenas and therefore to the researchers. It's a place where new cubs can be identified, mother/cub relations determined, and rank relationships worked out. A lot happens at a communal den, so it's the optimal place for hyena research. Mike had a hunch. The Talek hyenas had once established their communal den in a forest plantation just outside the reserve. He'd noticed a lot of hyenas coming from and going in that direction lately, in which case they might well have moved back to the old neighborhood. The plan was to go and take a look.

Arriving at the edge of a shallow ravine, Mike pulled the handbrake. He pointed to a wooded area farther along the far side and said, "That's the plantation over there. We can't get across to it from here, though." I was already pulling on the door handle as I said, Not in this thing we can't. Mike looked at me. "Are you serious?" Yes, I am, and you'd better come along just to make sure no hyenas eat me. I turned to Tigist in the backseat and asked if she wanted to come along. She didn't trust hyenas outside Harar; there was no way she was going to get out of a car in hyena country. Still, she trusted my judgment and didn't try to stop us when Mike and I got out of the car and finally stood on the hard ground of the Maasai Mara.

The first thing that struck me was the silence. No engine idling, no radio receiver, just the quiet of the landscape punctuated by the birds calling up the sunrise. The second thing was the exposure. As I slid down the bank into the ravine, I felt gloriously vulnerable. I imagined a leopard in the thicket beside the creek, tail flickering, paws twitching, ready to spring from the cover and dispatch me with a bite to the neck. There was as much chance of that happening as there was of meeting the king of England, but the fear and trepidation made me feel alive and in the company of my ancestors. Silent apart from our breathing, we walked along the sandy creek bed and came to a worn path on the bank with a lot of hyena tracks. Mike signaled for me to stay quiet as we climbed up the bank, because the den should be just on the other side. We cleared the top and arrived at a little grassy clearing.

At the same time that Mike pointed across the clearing to a depression, a hyena's head peered at us over the rise. She froze, her eyes went wide, and in an instant she was off, uttering alarm rattles as she ran. Instant pandemonium ensued. Hyenas appeared from everywhere and scattered across the clearing, disappearing into the forest with alarm calls snickering across the ground. Mike commented, "I guess we found the den." I was grateful. I felt bad about disturbing so many hyenas, but I realized what had been so lacking before: recognition. No longer an innocuous head peering out from inside a vehicle, I was on the ground, all too human, and all too presently terrifying. And so the hyenas responded.

Later in the morning we went with Steph to a grassy airstrip where a single-engine plane arrived to fly us back to Nairobi. From the air, the Maasai Mara looks a lot like parkland. The expanses of grass punctuated by ungulate-pruned trees, the various trails, and the absence of houses stand in marked contrast to the messy, crowded land outside, full of people, livestock, and the materials on which they depend. This is no accident of nature. The Maasai Mara really is a park. The boundaries are defined, there are gates at the entrances, you must pay an entrance fee to get in, and there are rules for maintaining the place in its parklike state. Within the boundaries, visitors are driven around in tour vehicles like Sunday picnickers on model railways. If someone spots a pride of lions sleeping under a tree, the vehicles converge on the spot and surround the lions, and everyone points a camera at the big cats, who are so accustomed to the attention they don't even wake up. And still no feet touch the ground, as though that somehow ensures that there is no human impact. The Maasai Mara is a built and carefully controlled environment.

Yet for all its defined boundaries and efforts to separate it from the locals, the Maasai Mara is very much frayed at the edges. Given the regular rainfall, there is plenty of graze to be found within the reserve, and many Maasai drive their herds into the park in the morning to fill up. Moreover, some of the carnivores spill over into Maasai lands and supplement their wild foods with some fattened domesticates. This can feed back into the park when a disgruntled farmer laces one of his dead animals with poison. A single hyena dying in the reserve with a belly full of poisoned goat can pass the stuff on to vultures, jackals, and other hyenas, and the effects just keep rolling on like the hills.[10] No matter how hard you try to create a natural environment, it's always going to be caught up in the social, economic, and political affairs of its creators.

To their credit, the hyena researchers in the Mara don't shy away from the unnatural side of this natural environment. While Joe Kolowski was following hyenas around at night, he noted their use of a refuse pit near one of the park's lodges. The pit was closed halfway through the study period, giving Kolowski a chance to see how the hyenas shifted their space use when the lodge's leftovers were taken off the menu.[11] Prior to that, Erin Boydston compared hyenas' space use over two periods separated by six years, during which livestock grazing increased in the park. She was interested in how the hyenas responded to human incursions; how did they alter their space use and activity patterns after Maasai pastoralists increasingly drove livestock into the middle of their territory? Biology will never be about organisms in "natural" environments. There will always be humans at the edges with their domesticates and poisons, humans in the middle poking cameras out of vehicles, and humans marking political and economic lines around green polygons on maps. For all its expanses of grass, the Maasai Mara is a very messy, tangled place where hyenas' ecologies, socialities, and histories are inextricably caught up with those of humans.

On flying out of Kenya, I had a new outlook. Whereas I'd previously seen myself as a hack anthropologist doing hack research on semidomesticated hack hyenas, I began to see my project in a different light. In the first place, the hyenas in Harar were not a bunch of second-rate, garbage-munching inferiors to their free-ranging predatory cousins in the Mara. Both groups lived in the shadow of humans and survived by dint of human indulgence—or human indifference. If anything, it was the Harar hyenas who deserved attention, because they represented the future. On a continent experiencing enormous population growth, unfettered development, and massive habitat destruction, it was the hyenas like those in Harar who stood a chance of persisting beyond the boundaries of protected areas and zoos.

I also came to feel good about my own practices. Whereas the Mara project was taking a top-down approach—collecting data to test hypotheses formed out of evolutionary theory—I was working from the bottom up. Rather than controlling for variables, I was exploring the limits of the variable in Harar, dissecting the interests and agendas of the various players in an urban human/hyena mixing bowl, and asking open-ended questions. What's important to a hyena person in a maze of narrow lanes? What's important to a human person encountering that hyena? What's it like to navigate lanes at night where you might encounter humans at every turn? I was exploring the possibilities that arise when hyenas have to coexist with the old evolutionary enemy, examining socially, politically, ecologically complex relations and how these two and other species operated to mutually shape human/hyena coexistence in a town of a hundred thousand people and two hundred hyenas. This is why it was so crucial for me to be crawling through hyena holes and following hyenas in the middle of the night. I needed to be there when the hyenas did things that expanded the realms of possibility. Epistemologically prepackaged, limited in presence, but unlimited in scope, I was on the ground, in the dark, and subjective as all hell. What better way to access the subjectivities of Others?

# 14

## TALKING UP HYENA REALITIES

I felt glad to be back with the hyenas in Harar, although I was spending more time with them at the garbage dump in the morning than in the lanes at night. It was a little safer. I was also thinking ahead to writing a dissertation. I had hundreds of hours of hyena observations, a stack of completed surveys from Harar and beyond, and hundreds of pages of notes from informal discussions. But as far as formal interviews went, and nice quotations that I could use to underline my findings, I had little. Returning to Harar after three months in Australia gave me a chance to rectify that, so I set about preparing for a series of formal interviews. I didn't imagine that this could be all that difficult, but then Harar has a way of making the simplest task into an epic personal struggle.

First, I enlisted my Harari friend Mahir as translator. Mahir was a tall, handsome chemical engineering student from Bahir Dar University who spoke excellent English. He was born and raised in Addis, which I saw as advantageous because, unlike a local, he wouldn't be so quick to bring his own beliefs about hyenas into his translations. Mahir would also recruit the interview participants, who would in turn be asked to recommend friends and relatives for additional interviews. In the social sciences this is known as snowball sampling, where you recruit subjects along the lines of social networks. It works pretty well in a town like Harar, where you have only two or three degrees of separation among the whole population.

At that stage I no longer had my office, so we did our first interviews at Ali Bal Café. Despite the café's state of disrepair, it was a popular place to meet for coffee. It was a nice neutral setting for interviews, too, although the noise of the

Bajaj taxis made recording difficult, and there was always some sort of distraction—local boys fighting over some insult, or the waiter chasing off urchins who poked their hands through the wrought-iron railing. The interviews went well. In general, people expressed positive attitudes toward hyenas. One Somali woman, originally from Jijiga, said that as a child she had learned to fear hyenas, but that since her first encounter with a hyena in Harar, she felt safe. One young Harari man said that the primary benefits of hyenas came from tourism. He insisted that the government should provide bones for the hyenas to eat because they generated tourist income. I'd expected these positive attitudes toward hyenas in Jugol; my surveys during Ramadan bore this out, and when you think about it—cleaning the streets, attracting tourists, keeping away jinn and dangerous hyenas—what's not to appreciate? But the reasons people gave for their positive views of hyenas stemmed from the tangible benefits hyenas provided. I was looking for someone to talk about the spiritual aspects of hyenas in greater depth, and I realized that this was unlikely among young people. Yet that was the demographic that was emerging from my snowball sampling. I needed to interview some old people.

I was apprehensive; I'd already banged into the recalcitrance of Harari old folks. They invariably canceled or cut short interviews, and even when they were inclined to talk, they often gave answers that were peripheral to the questions I was asking. What I realized was that I was up against an immovable institution in Harar, one in which old people can behave as they wish toward their juniors and are guaranteed respect regardless. Elders have psychological power over young people that is rooted in tradition and that gives them license to interrupt, push aside, ignore, abuse, and even physically beat their juniors with no consequences to themselves. Old equals automatic respect equals free rein. They don't have to be wise, they don't have to be polite, thoughtful, kind, or well-mannered; they just have to be old. The upshot is that old people feel no obligation at all to cooperate with a (comparatively) young anthropologist and his whippersnapper of a translator. As far as ethical research goes, it's close to perfect, because there can be no possibility that the subject feels coerced or obliged to cooperate. But in terms of gathering data, it's debilitating.

That month we made several attempts to interview old people, including visits to murids at shrines and visits to relatives who we thought might be more helpful. But in every case we found ourselves banging into walls. The most promising interviewee was Arif's aunt, who lived in Bedroberi. She was very welcoming and cooperative and even gave some interesting answers, albeit a little off topic. But after only a few minutes she got up and went to another room. Arif put his hands on his knees and said, "We can go now." Did I say something wrong? "No, she'd just had enough of the interview." I was incredulous. We were only just beginning to elicit some interesting lines of discussion when Arif's aunt abruptly turned off the tap. If I wasn't already frustrated, that was the sealer. I decided to forget about the interviews and go back to my hyenas. Willi and the others were far more helpful.

Fortunately, however, Mahir was determined to persist. He was making good money as a translator and, to his credit, he not only worked hard to find another interviewee but managed to keep me interested—at least for one last attempt. A young man named Muheddin—one of those we'd interviewed earlier—told his grandmother about my research and asked her if she'd like to talk to me. She said she would, and Mahir arranged a visit.

We met up with Muheddin at the main market and he escorted us through the narrow lanes to his house. It wasn't a traditional Harari house; rather, it was a double-story stone and mud construction with a tin roof.[1] We climbed up a precarious wooden staircase and entered through a door that bore the flaking remnants of dark green paint.

The woman who greeted us inside was dark and weathered. She was obviously not one of the wealthier Hararis, as her house and clothes were quite modest. She hadn't a colorful shawl, typical of Harari grandmas, or any jewelry on her fingers, just a plain green shawl wrapped tightly around her head in the Amhara style and clothes suited to domestic work. I guessed there were not going to be any servants inside. Mahir introduced her as Uma Halima (Grandma Halima). She smiled and bade us sit on the floor (there is no furniture in a burcha room) while she went through a doorway into a kitchenette. Mahir produced a bag of khat and distributed a few branches, and we made small talk while Uma Halima remained in the kitchenette.

After about ten minutes sitting and chewing khat, I began to wonder about the interview. Uma Halima was still busying herself in the kitchenette. I didn't want to be rude and ask if she would join us, so I asked Mahir if he knew what was going on. Did she know we'd scheduled an interview, or were they going to spring it on her in a few minutes? He told me to be patient, so I sat and chewed my khat, rendered helpless by my ignorance.

Another ten minutes passed, and just when I'd resigned myself to an afternoon's burcha and no interview, Uma Halima appeared in the doorway with an enamel tray. Muheddin quickly stood and received the tray, placing it on a plastic mat in the middle of the room among discarded khat leaves. It dawned on me: Uma Halima had made us lunch. We crowded around the tray, tearing off pieces of ukhat and scooping up *shiro*, the modest besan curry favored by Ethiopians of limited means. When Uma Halima disappeared back into the kitchenette, I decided that I was content with lunch and burcha, regardless of whether we held an interview.

After we finished lunch, Muheddin fetched a jug of water and a bowl so we could wash our hands. He held the jug above and the bowl below our hands, allowing a trickle of precious water to pass between as we washed. We retreated to our places, leaning against the living room walls, and resumed our burcha, talking a little less while we digested our food and munching on leaves like contented goats. Uma Halima returned to the burcha room, this time with a small table and a stool. She pulled the plastic mat back to the wall and set the table on it. The table had an understory with some small coffee cups; she took four and placed them on top, then set the stool against the wall. "How was the lunch?" she asked me in Harari. *Yittiman*, I answered. Delicious. She smiled and declared, "He speaks Harari!" I told her just a little, but she insisted that I must be fluent. Uma Halima ducked back into the kitchenette and returned with a black stoneware kettle, setting it on a little circular ring made from tightly wound grass. She took a seat and began pouring coffee.

Ethiopians consider making and serving coffee a solemn, ceremonious affair. It is both the task and the measure of a woman. In Harar, they make *du'a* before the coffee is served. This is a supplication, usually made in Arabic, asking God for a blessing. Once du'a is made, everyone relaxes and conversation is suspended while the woman pours and serves the brew. She takes the cups and hands them around to guests without a word from either giver or receiver. The first pour is called *abul*. It's the strongest. After abul is served, more water is added to the grounds in the kettle, which is returned to the coals to boil. The second serving is called *tona*. It's far more pleasant than abul, although Ethiopians practically solidify their coffee with bucketloads of sugar, so even the strongest is sickly sweet. After tona, more water, more brewing, and another pour called *baraka*. It means blessing.

We finished baraka and handed our cups back to Uma Halima. She set them out on the table, sat back on the little wooden stool with her back against the wall, and placed her hands in her lap. "You know how we make du'a for waraba?" Mahir was translating as she spoke. Seriously? "Yes," she said. "'Mountain hyena, may he find health, bones, and all his needs and then return to his cave.'[2] Forty years ago, everybody used to make du'a like this for the people and for the warabas; they used to pray for them. Of course you pray for yourself: 'Allah, give me food, give me what I need,' and that sort of thing. After that, before you finish, you make du'a for the waraba also: 'May Allah give food to the warabas.' But you don't hear this one now. The only place you will hear it is the shrines. They won't forget this du'a at the shrines because the murid will teach his children and it will follow for every child in his lineage. This will protect Harar, so nothing bad will happen between the warabas and the Harari people. Even if everybody in Harar forgets this thing, the shrines will keep it and pass it on."

Do the murids and hyenas have a connection?

"Yes, a special connection. The warabas protect the murids. If you think to yourself that you want to kill some murid or somebody there, the warabas will stop you on the road. They will make the oowhoop and they won't let you pass. If you try, they'll harm you. This is the connection. If the warabas are hungry, they'll go to the murid and say oowhoop. Now the murid knows they're hungry and tells people to provide food. The murid is looking after this waraba and the waraba after the murid. A favor for a favor. Say you make *mia*. This is a promise to visit a shrine and bring khat and things to the murid. This time the warabas protect you. You walk on the road to the shrine at night and the warabas come and walk beside you, two on the right and two on the left. When you get there, the murid will say, 'Were you afraid of this waraba? We sent him to accompany you so not a thing will happen to you on the way.' After that, you give the things and stay at the shrine for some time. When you go home, the warabas will go with you back to the town. I don't know why this is happening but it's still happening to people these days."

Listening to Uma Halima, I was reminded of a large female Aboker hyena I encountered in Jugol. I'd followed her down the hill from Jugol Hospital, where she nonchalantly passed through the bystanders like a human pedestrian. This drew not a few curious glances. The hyena went directly to an open gate, where she looked in at a celebration that was going on. When she nosed about just inside the gate, a man emerged, picked up a stone, and chased her off. I followed her along a lane to an intersection, where she stopped and turned to look at me. Figuring that I was in her way, I pressed against the wall and let her pass. Once I'd done that, she led me back across a metaled road and up some steep steps past the ancient mosque known as Dini. Without stopping, she took a zigzagging route through Assumberi, past Emir Abdullahi Mosque, down a narrow lane, and directly to my own gate. Ali, the clerk from Abdella Sherif Museum, later told me that the Aboker female was not only escorting me home. She was giving me a message from the other hyenas: they knew where I lived.

Uma Halima continued, "The first time I went to Abdul Qadir Jailan, it was night and warabas came on the road, two on the left and two on the right. I was so afraid, I sat on the road like this, crouching. The warabas sat on the road too and stared at me. Oh, I was so afraid, I said, 'Derma sheikh, derma sheikh, please go away.' This is what we say so they won't attack you, 'Derma sheikh, go peacefully,' and they leave you alone. But then I knew if the warabas are angry they put their tail up like this. If you throw stones at them and they straighten their tails, they will attack you. If their tail is down, then they won't harm you. So I got up and then the warabas went with me to the shrine. When I got to Abdul Qadir Jailan, the murid was there and he said, 'Tonight you came with our dog.' I realized the warabas were walking beside to protect me."

Uma Halima's summation of the hyenas' moods from their loosely hanging tails reflects a skillful attunement that people in Harar have to hyena body language—and vice versa. One night in Jugol, I was following the dominant Sofi male, Bouki, from Suqutatberi toward the main market. He'd already marched between two men standing on either side of the hyena hole, so I figured he was feeling confident. About halfway to the main market, he rounded a bend and went straight past a man sitting on a step with a bagful of khat. As Bouki's large, furry body passed before the man's face like a ship cruising along a narrow canal, the man snapped out of his ruminations and did a pronounced double take. Farther along the same lane, Bouki encountered another man coming toward him, carefree and swinging a bag of khat. The man glanced up and stopped dead when he saw Bouki. He jumped up onto a step and pinned himself against the wall as the big male hyena passed by. He'd immediately gauged Bouki's determination, and Bouki had gauged the man's lack thereof. Neither of them hesitated in what they were doing for even an instant as they read each other's intentions.

I looked at Mahir. He was concentrating hard on what Uma Halima was saying but at the same time a look of wonder was spreading across his face. Uma Halima was still talking. "When I was a child, you know when you are a child you have to usually get out your gate and play. So that the parents will trick you and tell you, 'There is waraba outside. Waraba is very hard, he will eat you, harm you.' Even then, we didn't know what he looked like, this waraba. Even as a child we didn't know this thing. Then one day my mother took me to market after evening prayers. We went to the market that night and when we were walking, there was this waraba in front of us. People said 'Waraba, waraba,' and I remember for me it was a dog. But when everybody said this, it came in my mind, everything I was told about waraba when I was growing up. I let go my mother and ran to the house crying to my grandfather and grandmother, 'There is waraba, waraba!' I said my mother was being eaten by waraba! My grandfather said, 'He won't harm you. From now on you know this waraba, so don't be afraid. He won't touch you.' I was seven or eight years old when my grandfather told me this, and after that I know this waraba won't hurt me because they have good relations with the people."

Children in Harar are fascinated by hyenas. Often, on encountering one in the lanes of Jugol, a boy will call "waraba!" and run after the hyena, who naturally runs in the opposite direction. In other cases boys feel compelled to throw rocks at hyenas, especially when encouraged by their friends. If no rocks are at hand, they might stamp their feet and try to scare them. Still, some children favor hyenas. Once, at the playing field, I watched as a man took meat scraps from a container and threw them to a waiting hyena. A pack of dogs came charging along and barked at the hyena, but she wasn't concerned. Then a boy picked up some rocks and threw them at the dogs to chase them away. The hyena watched closely what the boy was doing and then, unperturbed by the rocks whizzing past, turned her attention back to the food scraps.

Uma Halima went on, "One time some people started digging a deep hole on a farm, looking for water. They left it open overnight and a waraba fell in the hole and couldn't get out. In the morning, the people found him and made a ladder, but the waraba didn't understand what to do with the ladder. He just bit. After that, they pulled him out with a sling. He knew they were helping him, so he didn't touch anybody. After that, he got out. Then everybody was afraid to untie him from the sling. One man went to the waraba and said, 'I'm doing a good thing for you, I'm untying this thing, so don't think about hurting me.' Just like a human, he understood, so he watched the man untie it. The waraba was free, but it was daytime. He couldn't go home because people would chase him, so the man said, 'Go into the house and wait there.' He went into the house and stayed until night, when they let him go home. After some days or months, this man who freed the waraba lost a cow. That night the waraba came and oowhooped. A man who knew the language came and said that the cow was in such and such a place. 'They won't eat your cow, so come and get it.' So the man went and found his cow. A favor for a favor. This thing was told to me by my own grandmother."

Can you tell me about these people who know hyena language?

"In the night there is a sound the waraba makes. The person who knows the language counts the sound: five times, six times. After the sixth, the waraba will make slowly, ooooo.[3] This is the thing he understands. When there is someone who knows the language, the waraba will come to the house and say something. He makes the sound ten times or something and tells the person about his feelings. He wants the person to do something or tell somebody something. But the person must not tell that he knows the language or tell somebody how to understand. He must keep it because the warabas don't like people to talk about their language. If you know the language and you tell somebody, they'll be very angry. Maybe they won't attack you, but if you have a goat or sheep or cow, they'll come and eat it."

The haunting oowhoops that punctuate the nights in Harar are as intrinsic to Harar's character as the lonely fig trees that mark the locations of shrines. Loaded with mystique, the oowhoops sound back and forth across the town at night conveying messages like ships' semaphore—comprehensible only to the initiated. For the rest of us, we know something is being said, we just don't know what. On a June evening at the Sofi feeding place, a crowd of hyenas had gathered around Yusuf and some tourists. From the other side of Yusuf's compound, an oowhoop rang out and the feeding hyenas froze with their heads up. Another oowhoop and as one they abandoned Yusuf and the tourists and raced past the back of the compound in the direction of the dirt road. I ran around to see what was happening and found the hyenas converging beside the road at the bottom of the hill. Someone had dumped a donkey carcass there and the hyenas were head down, bum up, devouring the gift. Initially, there must have been only two hyenas, silently, secretly feasting on the donkey—silently, that is, until their excitement got the better of them and one displaced the other. At that, and in a bloody-minded fit of pique, the put-out hyena tattled to the rest of the clan.

Uma Halima was rinsing our cups and putting them back under the table. "Say some waraba comes from another place. This other waraba is very hard. You heard that in Harar there was something like a waraba [who] has bitten some child or hurt somebody. These are not Harari warabas. The one who knows waraba language will talk to the warabas. They tell him, 'We are not the ones who did this thing. Come and see; these are the ones, the other warabas. Kombolcha or Alemaya warabas are the ones who came to you and did this; we are not the ones.' The Harar warabas are very decent, they don't hurt people. The Harar waraba is protecting the people from this other waraba so they won't hurt anybody.

"Same thing with jinn. The warabas are good to stay in Harar because they will protect us from these jinn, they will eat these jinn. Waraba eats jinn, and after he eats he vomits. When he vomits, there are fingernails. You can take these and use them like medicine, use to protect against buda, jinn, everything. If you work and save some money, after that you could lose everything because somebody does something on you. This medicine will protect you from someone doing something spiritually against you, something like ordering jinn against you. If someone orders the jinn to do something on you, the waraba and the fingernails will protect you from this."

I remembered sitting at the shrine with Abbas one night, watching as Willi went across to his favorite place beside the canal. Willi tensed his body and started retching, repeatedly squeezing his abdomen between his ribs and his knees as though pumping something from his stomach. Before long, a slimy gag-ball emerged from his mouth and dropped onto the ground. Abbas thrust his hand in front of me and said with some urgency, "Marcos, flashlight!" I handed him my flashlight and he rushed over to where Willi was sniffing the product of his efforts. Willi turned and saw Abbas coming. He scampered off to a safe distance and turned to watch as Abbas knelt with the flashlight directed at the gag-ball, teasing it apart with a stick and examining the contents carefully.

Uma Halima had yet more to say about jinn. "My mother used to tell the kids, 'Don't open this door, don't let it open! Because maybe the waraba will chase a jinni into the compound and this jinni will tell you that I will give you anything—money, everything that you want—if you take the jinni to his home. But it's hard to deal with the jinn because they'll fool you. You have to know this Quran. Bismillahi. If they're going to your soul they will take you, be over you. The warabas will protect us from these things; they're good for Harar. The waraba won't hurt you. If you make a favor for him, he won't hurt you or any of your descendants or family. Nor will other warabas. He'll tell his whole family, 'Marcus did me some good,' so they'll protect you and your family and your animals."

Uma Halima stopped talking and looked at me. I was mesmerized. Mahir took a swig of water. He'd been translating rapid-fire for forty-five minutes, and it took a lot of concentration to listen to chunks of information and repeat them back to me in English. He looked at me. "Is there anything else you want to ask?" I was still absorbing the information and sorting it out in my head. I checked my voice recorder to make sure I'd recorded it all. It wouldn't have been uncharacteristic of me to forget to press the record button, and I dreaded the thought of asking Uma Halima to repeat everything she'd just told me. It was all there. I shrugged. Is it okay to use her words in publications? She replied, "Abshir." Relax, no problem. We stood up and thanked Uma Halima. She smiled graciously and saw us to the door, while Muheddin escorted us down the wooden steps and out of the quiet compound. As we stepped through the gate and out in the lane, two boys ran past, laughing hysterically while they guided their spinning metal hoops down the hill. We turned and walked back up to the main market in silence.

What was so captivating about Uma Halima's conceptions of hyenas was how they so elegantly encapsulated the majority view of people in Harar. Except for the du'a, very little that Uma Halima said was new to me. The stories of hyenas escorting people, returning favors, protecting murids, controlling jinn, and passing on messages to people with knowledge of hyena language were all things I'd been told in previous discussions with Hararis. But Uma Halima stood out not only for the coffee but for the way she so beautifully articulated everything that other people had told me in bits and strokes. Moreover, she brought to the table her own experiences with hyenas and her personal views about their intentions toward people. She essentially summed up what I'd been learning for eighteen months and put it all together in a concise, captivating lesson.

Still, Uma Halima's words present a challenge. How can her views be taken seriously without a patronizing nod and a wink? How can something so important to her be treated as anything more than a curious set of folk beliefs dreamed up during burcha? In the first place, the things Uma Halima told me are not just khat -inspired ruminations conjured out of dark, smoky rooms. They are empirically grounded ideas born of people's encounters with hyenas in the lanes of Jugol and precipitated in social interactions among people. When a hyena puts his nose to the ground outside someone's house and whoops, it really does appear as though he's passing on a message to someone within. And at the end, he really does make a long mournful ooooo. When a hyena plods along ahead of you, it really seems like he's escorting you. When hyenas vomit, the slivers of hard stuff in the slimy hair balls really do look like fingernails. Of course, a positivist could descend on Harar and test Uma Halima's views, and probably he would find other explanations. But is that the end point we're looking for?

When I describe the events of Kombolcha to people in Australia, I'm often asked if it is really the case that hyenas exact revenge on people for killing their clan mates. This fascination for the "really real" perturbs me, because it implies that we non-Hararis can access some kind of objective hyena reality at which the locals can only guess. When it comes to realities, though, we, like the Hararis, can only construct them. The biological sciences' behaviorally driven hyena is as much a construction as the Hararis' socially engaged actor-hyena; both are empirically grounded and both make unsubstantiated claims about hyena subjectivities. For Hararis, it is a given that hyenas are persons with mental states; for those in the biological sciences, the null hypothesis holds that mammals other than humans are not. In both cases, it is this subjective attribution or otherwise of personhood that underpins their policies toward hyenas. Yet the hard sciences' construction of hyenas predominates, not just because of academic rigor but because of overrepresentation and carefully constructed language that generates these realities, a language that mutes the subjectivities of hyenas and silences the voices of the locals. From there emerge globally generic environmental policies under which hyenas become natural resources that can be managed, relocated, and culled. Meanwhile, the Harari constructions of hyena subjectivities that engender policies of respect and genuine efforts to coexist are muted. Being as fond of hyenas as I am, I know which of the realities I prefer.

# 15

## LOOKING THROUGH A HYENA HOLE

The garbage dump just outside Suqutatberi became something of a regular haunt in terms of my social relations with the Sofi hyenas. I went there in the morning before sunrise and spent time with the hyenas, waiting for the garbage truck to arrive with dumpster loads of fresh rubbish. By now, the regulars at the dump were almost completely unconcerned with me. They lifted their heads and looked when I arrived at the top of the dirt road leading down to the mounds of rubbish, but as soon as they recognized me they returned their chins to their carpals and continued staring straight ahead. Only Willi paid me more heed than that. Once I found a relatively clear, dry patch of ground and sat, Willi got up and ambled over. He was like a child who greets you enthusiastically but doesn't know what to do after that. I usually offered him his comb, which he'd demolish with a few quick bites before nipping at my arm. I was buying combs regularly by this time, and the owner of the shop at Argobberi where I bought them must have thought me a strange farenji indeed, though he never questioned my comb fetish and reached automatically for the comb display whenever I walked into the shop.

My favorite hangout was a secluded clearing that lay between some farmland and a stream. I called this place "the waiting room" because hyenas congregated there in the morning and listened to the garbage truck while it made its rounds of the town. When the truck arrived at the dump, some hundred meters away, the hyenas sat up and listened intently to the garbage being unloaded. Sometimes they got up and went to investigate, while other times they lay back down and remained in the waiting room or else headed for home. I never worked out what differences they heard in the sounds of the garbage being unloaded, but they certainly were particular about which loads of garbage were worth investigating.

The waiting room was relaxing, mostly because there were no people there. We could lie on the grass while the rising sun burned off the morning dew and not worry about being disturbed. I remember Willi and Kamareya wrestling right beside me while I lay on the grass, Kamareya, not a meter away, turning his head to look at me before Willi bit his neck and drew him back into the game. My field notes from that time are sparse indeed. I was less observer than participant, and I was happy to lie on the grass with the other hyenas just listening to the sounds of morning. There was little worth recording anyway. The occasional hyena arrived and lay down by herself or went to sniff a clan mate; the occasional hyena got up and left. I expected from there that my remaining time with the hyenas would be uneventful, and I'd discover nothing new. Willi had other ideas.

One morning Willi and I were wrestling under a habuka tree while the eastern sky turned a misty pink, heralding the coming sunrise.[1] I decided to leave the garbage dump and head out into the farmland, so I got up and went along the trail that led across the road to the waiting room. Willi followed, and we arrived together at the top of a steep slope that led down to the dirt road. The slope was always a challenge for me, but Willi showed me a new way down that looked easier. He dropped from point to point like a mountain goat while I slid down the dirt and broke into a run at the bottom, slowing myself up at the ascent on the other side of the road. We climbed the opposite side together—this time I was more capable than Willi—and breached the hedgerow, to find some other hyenas in the waiting room. Hadha Kamar was there, looking at us through her sultry eyelashes. Chaltu and Muka Dibbey were lying in the sun on the far side of the clearing. Willi and I lay down near Hadha Kamar and listened for the garbage truck. I could see that Willi had picked up the sound, but he wasn't listening intently. Only when the truck came rattling down Jugol's outer road did the hyenas lift their heads. By this time I could hear it too; when it arrived at the dump we all listened to the hydraulics as the mechanical arm lifted the dumpster from the truck bed and tipped the contents onto the ground. I imagined the children there, converging on the fresh garbage, seeking anything that could be used or traded. Old plastic bottles were prized in this land of limited water storage.

The hyenas made a collective decision about the garbage. It was too late in the morning? There were too many people? There weren't enough kilos of bones per ton? I had no idea. But the decision was made, and I watched their spotted rumps amble off into the light of the early morning sun. Willi was the last to go, and he looked over his shoulder at me, inviting me to follow. I had no intentions; I just wanted to follow as far as I could before I lost him, so I trailed after as he led the way. He pushed through a hole in the thorn hedge that lined the waiting room. It was a hole I had no chance of getting through without being shredded, so I took the long way around to where the hedge was low enough for me to hop over without being snagged. I ran to catch up with Willi, but he was taking his time.

He led me along a path to another thorny hedge. There was yet another small hole in the hedge, but this time Willi didn't push through. He looked at me and then trotted along the hedgerow before seeming to disappear into it. I thought I'd lost him, but when I arrived at the place where he'd disappeared, I found a nice wide opening, wide enough to allow me through with only a few scratches. I held my arms close to my chest and pushed through, to find Willi waiting on the other side. He turned and made off toward a creek crossing, occasionally turning to check on me as he went. When he broke into a canter, I wondered if he'd had enough of my following, but he pulled up and waited for me to get closer before setting off again. I felt like I was following Lassie to the old mill where young Jimmy was trapped under a roof beam.

The sun was rising higher, and I imagined the landscape would soon be crowded with farmers heading out to tend their crops, their munchas conveniently at hand should they need to cut down a fleeing hyena. We arrived at a place where an irrigation channel bridged one of the streams that ran east from Jugol—water crossing water. The channel had broad edges, so it was possible to cross, but Willi was reticent. He halted at the crossing and doubled back to a place between two bushes. I had no idea what was worrying him, so I went and stood beside the channel and encouraged him. Willi emerged from between the bushes and began to cross, but when he was about halfway across we both caught the movement of a human at the top of the rise on the other side. Willi dived off the crossing into the brush beside the stream below and raced along the stream bed, disappearing into the vegetation. I was startled but held my ground as the farmer approached to cross. I said a sheepish good morning in Oromo to the farmer as he came down from the top of the rise and crossed the stream. As he went on his way, he glanced over his shoulder with a look of suspicion.

Willi had disappeared, but I knew he was across the stream and that I was still between him and the Sofi dens. I figured that he'd have to pass me at some point on his way home, so I climbed the rise and sat down on the bare earth at the top. Sure enough, Willi emerged from a copse of eucalypts and trotted straight toward me. I held out my arm so he could have a reassuring nibble, and with a few tail swishes we were off once again, this time hurrying across some open ground to a place where we rejoined the channel as it emerged from some trees. This channel had no water, but the bottom was lined with damp sand that recorded

Willi's footprints as he trotted off, ducking under some tree limbs and disappearing around a bend. I figured he must have been close to his den, so I let him go and sat down on the side of the irrigation channel. But Willi wasn't done with me. He reappeared and crawled under the limbs toward me, dislodging a small branch, which brushed against his back. This gave him the fright of his life and he shot out from under the branch with his tail between his legs. He turned to confront his unseen assailant, at which point he realized that it was a branch. Nevertheless, he sniffed and nudged it with his nose to be certain that it was inanimate.

Willi looked at me before turning and ducking under the limbs to the other side, where he did an about-face and waited for me. I accepted his invitation and clambered over the obstacle, struggling and snagging my clothes as I went. Willi watched me impassively. I jumped down onto the damp sand, whereupon he turned and we were on our way again. He climbed a shallow bank and led me downhill across a little clearing to where a wall of rushes and reeds hid the stream that we had crossed earlier. I sensed that we were now very close to his den. We followed a well-worn trail to a little archway among the reeds, a hyena-sized entrance to a series of tunnels that constituted Willi's daytime home. I felt so privileged, and not a little unworthy, to have been shown this secret place that protected the hyenas from the terrors of the diurnal human landscape.

As Willi disappeared under the arch, I sat on the bank opposite, listening to the rustling within as he made himself at home. Willi went silent and then re-emerged at the entrance. He looked at me, turned, and disappeared again. I contemplated following. He was obviously inviting me in, and I felt a mixture of curiosity and obligation. He wanted my company, and a refusal on my part amounted to a snubbing. But I was torn. There were probably other hyenas in there who would not be as enthusiastic about my visiting the den. I imagined my appearance inciting mothers to abandon cubs, or worse, defend them. I also imagined the kind of sediment that washed down from Jugol and covered the floor of the dens with each flood, the floor I'd have to crawl over on my hands and knees. In the end, my imagination got the better of me and I remained where I sat, content with the morning's discovery.

But Willi was not yet finished with his efforts. He emerged again at the tunnel entrance, walked straight to me, and tried to bite. I yanked a dry cornstalk from the ground and offered him one end, which he took in his mouth. We had a little tug-of-war over the cornstalk as he tried to drag me toward the entrance to the den. I let go of the cornstalk and he took it with him toward the den, looking over his shoulder as he disappeared once again into the hole. I decided that I should leave him in peace, so I stood and followed the path back to the open country, from which I could see Jugol's turquoise facets shining in the morning sun. I marched uphill through the cemetery and had broken into a sweat by the time I reached the Argobberi gate. Followed by a few curious gazes, I disappeared into the maze of Jugol's lanes.

In a research facility in Berkeley, California, a female hyena named Nakuru enters an enclosure. A human observer outside the enclosure takes notes, another two operate wall-mounted apparatuses that hang above Nakuru's head, and a video camera mounted in the corner records the scene. An automatic gate opens and a male hyena, Winnie, is let into the enclosure. Winnie approaches Nakuru with his tail between his legs, his head held sideways, and his teeth bared. Nakuru pushes her snout under Winnie and sniffs. Winnie responds before the two disengage, swishing their tails as they go. Jerking his head a little, Winnie investigates the enclosure. As he sniffs at the ground beneath both apparatuses, Nakuru watches him intently. There are yellow ropes hanging above the hyenas' heads, two for each apparatus. When Winnie reaches up and bites on one of the ropes just above the end, Nakuru joins him under the apparatus and bites on a second rope. Winnie pulls hard on his rope, Nakuru pulls on hers, and a platform above them tilts, allowing a pork bone and a ball of cat food to drop on the floor between the two hyenas. They eat.

These events come from an experiment conducted in 2008 at the University of California, Berkeley.[2] Over several trials of three different experiments, eight captive-raised hyenas were tested, and all demonstrated a natural ability to cooperate to solve problems, something that had never been tested in hyenas before. This piques my interest, and not just because a bunch of hyenas demonstrated that they could team up to open a trapdoor and get food. When I dwell on it, it gives me a way in; it offers me a rope that, if I choose to pull, opens up my relations with Willi. Despite its implied replicability, the Berkeley experiment is in fact fully loaded with one-offs. There are messy histories, uncertain futures, acts of sex and violence, declarations of love, and tendrils of complex relations that reach far and wide. In which case, we find a lot more going on than two hyenas pulling on a couple of ropes, and a lot more at stake than a pork bone and a ball of cat food. There are cubs taken from dens in the Maasai Mara, gonadectomies, betrayals by film companies, human/hyena bonds, feminist critiques, zoo placements, animal rights protests, and continuous efforts to raise funds to maintain the colony.[3] There is even a farmer somewhere whose chicken by-product constitutes the greater portion of the prepackaged Hills Prescription Feline Diet the hyenas are eating. We need to ask what is at stake here. Who stands to gain, and who to lose? Who is already benefiting, and who is losing? If the hyenas don't cooperate to access the food, will the results be published? Since they did cooperate, will the findings be used in applications for more funding? How will the performances of the hyenas affect their future? The published findings compare the hyenas to primates for their cooperative problem-solving skills, outperforming "well-trained" chimpanzees on synchrony and spatial tasks. What does this mean for hyenas in the wild? Cooperative problem solving is something at which humans excel. Given that hyenas will never measure up to the human standard, is it actually a form of oppression to test the (limited) degree to which hyenas compare?[4]

This experiment also reveals some unexpected power relations. When the researchers set up the apparatuses with the food rewards and lengths of rope, a lot of human interest rode on those two hyenas. It's a kind of disempowerment when so much is at stake, and all in the hands (or paws) of the hyena research subjects. And this brings me back to Winnie and Nakuru and how the experiment was set up. During the baseline tests, the apparatus delivered food at the pull of a single rope. In this way, all participating hyenas were given experience in pulling a rope and being rewarded with food. But in the test trial, two ropes had to be pulled simultaneously on one of two available apparatuses, and only Nakuru had knowledge and experience of this. Winnie was naïve, so Nakuru had to monitor him and join him at whichever apparatus he chose. When she did, she had to pull the second rope in synchrony. That's pretty smart, but that's not what interests me most. Nakuru also outranked the inexperienced Winnie. After several trials during which Winnie worked out how to pull in synchrony, Nakuru began taking the lead, and the previously naïve Winnie followed her to whichever apparatus she chose. But in the first few trials, Nakuru had no choice. If she wanted to eat, she had to monitor her subordinate, follow his choice of apparatus, and synchronize her rope pulling with his. In this world of inter-/intraspecies relations, knowledge can be the inverse of power.

The Berkeley experiment shows me a way in to thinking about what was at stake in Harar in my relationship with Willi. In coming together, we put down roots into histories, prehistories, colonizations, landscapes, ecologies, and global processes, which in turn fed complexity and messy relations of power back into our nightly encounters. And the deeper the digging, the messier it gets.

When I arrived in Harar, I carried with me a lot more than a bag full of hyena-watching gear. I brought my evolutionary baggage: my all too human fascination with large carnivores and my hominin bipedality, which puts any hyena on edge. As a middle-class Australian boy, I brought with me Disneyish fantasies of wild-animal friends;[5] as an avid reader of Paul Shepard, I brought a skepticism about such things and a directive to keep the Others at a distance; from my code of ethics for wildlife research, I brought my mantra: Don't disturb the hyenas! That's a little difficult when you're a human on foot in a narrow lane. From my supervisor, Debbie Rose, I brought ideas about the research process as a mode of openness to possibilities, and from my archaeology background I brought ideas about needing hypotheses to test. As for Willi, he brought with him the evolutionary baggage that fostered mistrust of humans, but also a savvy, hyaenid opportunism that allowed his forebears to persist in a landscape crowded with humans and their domesticates. The locals' beliefs and their reliance on khat over pastoralism fostered the persistence of hyenas. In a landscape teeming with khat thieves, dik-diks, and jinn, hyenas are good to have around. In times of famine and war, which are disturbingly commonplace in Ethiopia, the hyenas thrive; in times of peace and prosperity, they tread water, eating mostly skin, bones, entrails, and donkeys, the last carrying a biblical prohibition against consumption by humans.[6] Harar, being a densely populated place full of meat eaters, became a magnet for hyenas, and throwing scraps complemented a growing tourist industry. Willi inherited the boldness of his relatives, and that combined with his youthful curiosity and openness to possibilities fostered by a historically informed feeding place. He quickly adapted to the crowded food-getting paradigm and extended his ideas out onto the roadside, where he first encountered me.

Given a heftier budget, I might have bought a research vehicle, although the narrow lanes of the town would have rendered it locally useless for following hyenas. I have the Oromo to thank for that. They occupied the region and compelled the Hararis to build a wall around their town, consigning the town to internal complexity. But money also buys dart guns and GPS collars, which could have given me a broader perspective and saved me all that time following hyenas around Jugol. Instead, I had to be on the ground and do the following and mapping firsthand. My budget also influenced my choice of feeding places. If money hadn't been an issue, I might have paid Mulugayta for a year's worth of nightly visits and focused on the Aboker hyenas, with their playing field and easily accessible dens. Instead, I went for the cheaper option, and the one that didn't offend my middle-class, Disneyish fantasies. Like the tourists who needed to see hyenas, the guides who needed to satisfy tourists, and the hyena men who needed to provide hyenas, I needed to make hyena observations within a limited budget and time frame in which to write a dissertation. Moreover, the dissertation had at least to be interesting, so when Willi presented himself to me and began sniffing my knee, I let it happen and quickly scribbled it in my notebook, knowing full well that it made for interesting data.

This is where power relations between species get messy. I needed Willi to do things. But, like Nakuru the hyena at Berkeley, my greater knowledge and experience did not automatically equal greater power. I had to sit, wait, monitor, get up, follow, and be in the right place at the right time when Willi did interesting things. I needed him to do interesting things. But my hanging around in the right place at the right time was not Willi's choice. This is why relations are so precarious: because nobody is really in control. There are imbalances in knowledge, disparities in awareness, and, just as important, different needs. Certainly, our needs can lead us to coerce and cajole—like my encouraging Willi to cross a stream, or Guramurey's waiting for Yusuf to throw meat beyond the lamplight—but where there is need, there is a scope for powerlessness and for the subsumption of both parties by overarching processes. Evolution, food competition, wildlife tourism, and academic research, among other things, all fed into and off the needs of Willi and myself, wresting control of our relations from both of us. Where various factors impelled me to make observations of Willi, I was compelled to follow his choices of lanes and his schedule. But if Willi needed to double back and go the other way up a lane, this caused a localized power shift. He turned and looked up at me, expressing his need that I move out of the way and let him pass. Within relations between species there is a constant to-and-fro of needs from which power and powerlessness emerge in some unexpected places. A human or hyena person brings a set of needs within an ecological, historical, social, and economic framework and depends on the Other to pull the rope.

In this way, through a need for and reliance on the Other, the self opens up to personal transformation. I need Martin Buber back at the table here, because his I-You relation is so crucial to what goes on when human and nonhuman subjects come together. In order for a human or hyena person to give himself over to the actions of the Other, he must recognize the Other as a capitalized You. Only through the I-You relation can the I pause and wait for the You to fulfill his needs. The human or nonhuman You is empowered here, because his subjectivity is thrust upon the I who must wait for the You to act. In this way—in giving the self over to actions of the Other—the I becomes transformed. The I must go toward the Other's state of being, not only to be in the right place at the right time, but to apprehend the Other's subjectivity on which the self's needs depend.

This is something in the research process that I never anticipated and that only became clear to me in hindsight.[7] Through putting myself in Willi's path and subjecting myself to his choices and actions, I transformed myself personally toward a hyenalike state. By the time Willi arrived on the scene, I was already following the hyenas' schedules, arriving at the feeding place in the evenings, waiting on the hill for tourists to arrive, and wandering over to the feeding when they did. Then, in spending time with Willi, I was compelled to understand what was important to him. I saw the attraction in sick camels, the opportunities at the garbage dump, the disquiet produced by human voices at the top of a lane, and the peace and security of the patch of ground I called the waiting room. I learned to despise dogs and found humans increasingly disturbing. I felt secure in the darkness of power outages and exposed in the daylight. And when lying in silence among the Sofi hyenas on the hill, I found a profound feeling of belonging. In giving my self over to Willi's personhood, in apprehending him as a capitalized You, I underwent personal transformation in the direction of his subjectivity, to the point where I found myself sitting in front of a hyena den seriously contemplating going in!

For his part, Willi was becoming more like a human. In addition to consuming human foodstuffs and chewing on the manufactured bits and pieces of human society, Willi moved dramatically toward a humanlike habitus. Compared to most hyenas, his flight distance from humans was miniscule. He became more and more at ease in the company of humans, especially me. Moreover, he strove to communicate with me at a level that was human-friendly, and he tried so hard to understand what it was that I, as a human, was on about. For us, intersubjectivity was a two-way street, and Willi was meeting me in the middle.

For both of us, this intersubjective self-transformation involved divesting ourselves of much of our (pre)histories and then going toward each other. I didn't become hyenalike in a strict sense; rather, I became human-becoming-like-hyena through apprehending Willi's personhood, putting aside my evolutionary baggage and altering myself toward Willi's Otherness. He did likewise, dropping his instinctive fear and apprehending me as a subject of significance toward whom he altered himself to be hyena-becoming-like-human. In both cases, it involved a step away from our species potentialities, grounded in millions of years of antagonism, and a readiness to apprehend the Other as a capitalized You, an openness to the demands and potentialities that the face of the Other presents.[8] And in that intangible You-saying place of interspecies intersubjectivity, where our selves engaged in a going-toward each other, there lay the organic, amorphous fabric from which our persons were continuously, mutually, uncontrollably shaped.

I should note here that this untidy little package of interspecies co-created habitus presents a challenge for the ethnographer. If we choose to heed Barbara Noske's call and apply an intersubjective approach to the study of animals, we find the same issues arising that all ethnographers face everywhere: no one enters the field without affecting and being affected by her subjects. A hyena's adaptive fear of humans is exactly that: adaptive. It serves a hyena well to react with unabashed terror to the sight of a hairless ape walking about on two legs. We humans are hazardous to hyena health. So I can only wonder at the negative consequences of my engagement with Willi and how it engendered a very unhyenalike audacity in the presence of humans. How would his becoming-like-human affect his approach to other humans in a land where they don't all subscribe to the benefits of having hyenas around? It makes for a difficult balancing act, one in which the imperative to represent animals using anthropology's intersubjective approach inevitably entails potentially detrimental personal transformation for the nonhuman subjects who engage in the I-You relation with ethnographers. It's a conundrum that won't go away, because intersubjectivity and personal transformation are inextricably linked. In honoring nonhuman persons by attending to their subjectivities, we ethnographers are very likely going to foster transformations that may or may not be to the betterment of our subjects. It's a prickly path, but one from which we cannot turn away without turning away from the Others.

Finally, as I've demonstrated, it is not just the nonhuman subjects who undergo transformation through this kind of intersubjective going-toward. The ethnographer needs to be prepared for unanticipated, sometimes profound personal change that emerges from entering into nonhuman subjectivities and that persists beyond the ethnographer's time in the field. Subsistence hunters who mimic their prey in order to capture and kill often voice serious concern about adopting the perspective of the nonhuman Other. In accessing animal subjectivities, human persons risk going too far over to the other side. At stake is their very humanity; they risk being absorbed into the subjective Other-worlds of their prey and never coming back.[9] As a remedy, hunters set about rehumanizing themselves after a hunt—engaging with human others in protracted conversation and overindulging in human foods and practices.

Toward the end of my time in Harar, I found humans increasingly disturbing, especially in daylight, and have since taken pains to avoid crowds of people. Even small-scale social engagements put me on edge. Whereas coming home from work at night along deserted streets should be at least a little scary, I find it instead relaxing to be in the dark and immune to the human gaze. Beyond that, I've tended toward a hyenalike pragmatism, an easy detachment in which I consider just being with someone a very affectionate act. Sentimentality eludes me, to the point where I find letting go of things all too easy. And as for dogs, I cannot describe my feelings about those creatures without using expletives. To her credit, Tigist is very understanding of her odd, hyenalike husband. If anything, engaging with me at such a close, intersubjective level has brought about a pragmatic, anthropophobic hyenalikeness in her. It makes me wonder how far the ripples from Willi's pulling me into the hyena world can possibly spread.

My last week in Harar was unsettling. Hashim Idris, the man whose tailor shop I rented for my office, was shot. He had a heated argument with two land speculators in his office at the Department of Land and things came to blows. That night, the two men went to his house above my old office and fired shots at Hashim's door. Thankfully, he and his family were not home, but that was not the end of the matter. The next day, the two waited for Hashim to arrive at Feres Magala after work and opened fire in the crowded square. After the previous night's shooting, Hashim had armed himself, so he returned fire. This happened directly in front of the police station, but the police were too concerned with self-preservation to get involved. No point in getting killed for forty dollars a month. Hashim had taken cover behind a Bajaj, and one of the men came up from behind and shot him point blank in the back of the head, blowing out the lower half of his face. He was taken to a hospital in Addis but was too badly hurt and died soon thereafter.

The following few days were tumultuous. Hashim was iconic in Harar for his combination of fearlessness and integrity. The locals were outraged that he'd been killed, and large numbers of people turned out to protest at the government offices. The two land speculators responsible for his death were Oromo, and this fired up some considerable resentment between the Hararis and the Oromo. Some people broke into and wrecked the office of the Oromo People's Democratic Organization, and one official suspected of being in collusion with the killers was beaten. I got caught up in one protest on my way home from the market. I emerged from a lane and ran headlong into a large crowd coming up Andenya Menged toward the government offices. As I ducked back down the lane, the police arrived and fired tear gas into the crowd. That was enough to send everyone scattering into the lanes, which quickly filled with coughing and spluttering people. I took refuge in the pharmacy that belonged to my friend Mohammed Abdul. He'd seen it all before; he told me, "It's just tear gas. Close the doors behind you and you can go out through the back." But, for me, Harar seemed to be falling apart. The "City of Peace" was failing its title. I was concerned for Tigist, and we were both ready to get out of town. Thankfully, our visa had come through two weeks earlier; we had our flights booked, and it was just a matter of riding out the storm until our departure date.

The night before we left Harar, we went to the feeding place as usual. Yusuf was already feeding the hyenas when we arrived, so I sat down in front of the shrine and took stock of who was present. Willi was there. He glanced sideways at me to acknowledge my arrival before returning his attention to the feeding. He was busy circling around Yusuf, trying to keep Dibbey on the other side. At the time, Dibbey was harassing him severely; she even inflicted a deep bite wound on his rump that exposed the red flesh for days. Indeed, so intent was she on attacking Willi that she rarely noticed the skinny Tukwondilli following her around. Fintamurey was there as well. I figured that her mother must have outranked Dibbey, because Fintamurey stood right in front of Yusuf with her head in the feed bucket, while Dibbey waited expectantly behind. It's so strange how hyenas respect rank over size and aggression.

Ajebbo emerged from the house and demanded that I come inside. In the living room, Tigist was handing out presents and chatting with Naime. Yusuf's eldest daughter, Ardalle, was close to tears. She absolutely idolized Tigist, and though she might not have been terribly upset that I was leaving, she was devastated at losing Tigist. We sat and talked for some time before Yusuf arrived at the doorway with the empty food bucket. He'd finished feeding and was waiting for Nasruti to fetch a jug and some water so he could wash the flesh, fat, and dirt from his hands. This was my cue to go out and visit the hyenas, so I put my shoes on and went out into the darkness.

Over by the drainage channel where Willi should have been, I found Dibbey. She looked placid and relaxed, even though Tukwondilli was hovering a few meters away. I went over to the back of the feeding place and scanned the hill with my night-vision scope. There was a juvenile lying in a depression and a big female at the far side sniffing the weeds, but no Willi. I stood there in the darkness searching for a feeling. I wanted to feel my heart breaking at having to leave the hyenas; I wanted to feel robbed of a chance to say good-bye to Willi; I wanted to feel a sense of being torn away from the place where I'd found belonging with a hyena clan. But nothing. All I could summon were meta-feelings: a sense of loss at not feeling anything; a yearning for some kind of sadness that wasn't there. Kamareya had been missing for two weeks, Baby was long gone, Willi had walked away on my hopes of saying good-bye, and all I had was a sense that something within me was missing and a melancholy over its absence. If I'd had a tail I would have swished it. With hyenalike pragmatism, I moved on and found Fintamurey beside the fence. I had very little fondness for Fintamurey; she was arrogant, pushy, and disregarding of others. I sat down beside her, produced the comb that I'd intended to leave with Willi, and offered it to her to chew. She sniffed it, took it from my hand, and then dropped it on the ground, showing no further interest. That was the thing about Willi: he was interested; he wanted to connect, although a lot of the time he struggled to find a way in which to do so. I rubbed the fur between Fintamurey's ears and went back into the house, where I found Ardalle hugging Tigist, the tears streaming down her face.

# NOTES

INTRODUCTION

1. Kruuk 1972, 144.
2. Even where livestock predation is an issue, people can have positive attitudes toward predators. For example, lions and hyenas both prey on Maasai livestock, yet a study by Goldman, Roque de Pinho, and Perry (2010) found that the Maasai had positive attitudes toward lions. This contrasted with their attitudes toward hyenas, which were largely scorn, disgust, and hatred.
3. See, for example, Glickman 1995 and Brottman 2012. Both authors try to understand why hyenas are so vilified, while people who work closely with them are so fond of the creatures.
4. See, for example, Louchart et al. 2009. It's well-nigh impossible to tell from bone fragments whether ancestral hominins were victims of hyena predation or whether they were scavenged. But hyenas are predators who prey on small children, so it's a pretty safe assumption that they used to prey on our ancestors.
5. Kruuk 2002 makes this point.
6. Of course, the footprints refer to those of Laetoli (Leakey and Hay 1979). A Musukuma tracker identified the prints of four hyenas at the site in addition to three hominins. The most common carnivores in the assemblages at Laetoli were hyenas (Barry 1987).
7. In my dissertation (Baynes-Rock 2012), I criticized the assumptions on which this debate was based, the foremost of which was that relationships between humans and other animals were predictable and consistent over time. From that assumption, this assemblage became the grounds on which our ancestors were determined to be either hunters or scavengers. See, for example, Bunn 1981; Potts and Shipman 1981; Shipman 1983; Binford 1985; and Blumenschine 1987. Relations between contemporary hyenas and humans are entirely contingent on things like time of day, hunger, and numbers of individuals, so it's a pretty long bow to draw, saying that the lifeways of our ancestors (and there's no evidence that the individuals at Olduvai were our direct ancestors) can be read from a single assemblage.
8. Quammen 2003, 328.
9. Morwood and Jungers (2009) argue for an earlier dispersal of australopithecines based on morphological similarities between australopithecines and *Homo floresiensis* found in Flores, Indonesia.
10. Based on findings from Spain, Italy, Georgia, and China, Arribas and Palmqvist (1999) argue for a dispersal of *Homo* from Africa to Eurasia in the Lower Pleistocene. This, according to the authors, was concurrent with a dispersal of saber-toothed cats and giant hyenas, which suggests that the three species were ecologically related.
11. This distribution is based on the occurrence of hyena fossils across Eurasia. See Gibert at al. 1999; Boaz et al. 2000; Gabunia et al. 2000;

Schepartz, Miller-Antonio, and Bakken 2000; Gibbons 2005. The fossils are found all too commonly with those of *Homo*.

12. I'm referring to a beautiful antler sculpture of a hyena from a cave at La Madeleine (Kurtén 1968) and an ochre painting in Grotte de Chauvet of a single spotted hyena (Chauvet, Deschamps, and Hillaire 1996). Paleolithic representations of hyenas are very rare in comparison to horses, bison, and even lions.

13. See Stiner 2004. I'm not sure that competition with wolves is sufficient reason for the disappearance of hyenas, considering that hyenas are well adapted to woodland environments in Africa.

14. Garrard 1984 argues that direct competition with humans over resources was the reason for hyenas' disappearance in the Levant after 10,500 B.P. The dwarf species, according to Rabinovich 2002, are *Crocuta crocuta debilis* and *Crocuta crocuta dorotheae*. The latter had reduced carnassials and larger premolars, indicating a heavier reliance on bone than flesh in comparison to their larger cousins.

15. See Mills and Hofer 1998. The evidence for reduced distribution of spotted hyenas is a comparison between historical and contemporary records.

16. I recognize that the term "witch" is loaded with cultural biases and preconceptions (Ciekawy and Geshiere 1998). However, this is the term that is common in literature about the kinds of people in Africa suspected of using supernatural means to inflict harm upon and/or consume the souls of others; see, for example, Johnstone 1902; Austen 1993; and Morris 2000a.

17. The hyena/witch association is widespread in Africa, from Ethiopia, to Malawi, to Senegal. See, for example, Beidelman 1963; Gray 1963; Niehaus 1995; Rutherford 1999; and Sapir 1981.

18. Accounts of exposure burials come from Routledge and Routledge 1968; Leakey 1977; and Gottlieb 1989. I suspect that this concern is no longer widespread, as missionaries and governments discourage exposure burials. Hyena predation on humans potentially perpetuates this (added) danger.

19. Accounts of hyena predation in Africa are abundant. See, for example, Sayers 1930; Balestra 1962; Kruuk 1968; Long 1973; Marshall 1976; Morris 2000b, and numerous others.

20. Scores of African folktales portray the hyena as the trickster who always ends up in the soup. For some examples, see Beidelman 1980.

21. Hemingway 1935, 38. Hemingway's account reads as a comment more on his guide than on hyenas.

22. Kruuk makes reference to Harar both in his book about spotted hyenas and in a *National Geographic* article (Kruuk 1968), which is aimed at dispelling myths about spotted hyenas as scavengers.

23. Burton 1856, 321. There is no indication that Burton actually witnessed this, and if it was indeed occurring then, it had to have been infrequent. Hyenas give birth to small litters after a long gestation, and the cubs have long weaning periods; it would have been a struggle to provide the numbers required for frequent spearings.

24. Paulitschke 1888, 142. Again, it is difficult to disentangle the fancy of the writer from what was happening on the ground. If the Harari elite ever hunted hyenas, this practice was certainly no longer pursued at the time of my visit.

25. Leslau 1965, 24.

26. Gibb 1996, 191. Gibb was primarily concerned with how the Harari people, the *ge usu*, maintained preeminence over other ethnic groups through inclusivity. Her accounts of hyenas in Harar are incidental but invaluable.

27. Rough terrain also prevented Kolowski et al. (2007) from following one of the clans in the Maasai Mara.

28. Marshall Thomas said her methods were unscientific. This is in fact key to the way she accessed the subjectivities of her dogs.

29. I later replaced the SLR and lenses with a point-and-shoot camera that was easier to carry around.

30. See also Nelson 1993, 217, which describes how the Koyukon of northern Alaska saw radio collars as "dangerously offensive" to the animals who were made to wear them.

CHAPTER 1

1. As in India, the locals in Harar sometimes refer to these taxis as "tuk-tuks." Normally, though, they call them Bajaj, which is in fact a specific Indian manufacturer of three-wheelers and motorcycles. The loud diesel models are made by another Indian manufacturer called Force.

2. See Littman 1936. Littman provides no sources for this information.

3. Records are uncertain on this, and others have suggested to me that Harar was an emirate

long before Abadir. There is a list of Harar's emirs in the Arthur Rimbaud Museum, which has Abadir succeeding Emir Haboba in A.D. 999. Gibb 1996 cites an identical list from the Harari Cultural Museum.

4. Hararis of the diaspora also celebrate Abadir at community gatherings. An outsider might say that this is flirting dangerously with idolatry, but Hararis insist that celebrating Abadir is markedly different from idolizing him. Certainly, there are no images of Abadir.

5. Actually, *farenji* is the more common term in Ethiopia (plural: *farenjis*). The Hararis and Oromo in Hararge say *farenjo*, as it's common for them to add an "o" to a name as a form of endearment. To confuse matters, the Amhara in Harar say *farenji*, and so do many educated Hararis.

6. Levine 1974 argues that the Oromo system of age sets, known as *gada*, was crucial to Oromo expansion. Specifically, it was the decentralized government and the tradition of attacking neighboring villages that allowed the Oromo to fill the power vacuum created after years of war between the Ethiopian Muslim and Christian states.

7. Trimingham 1965, 187. Trimingham's description elegantly captures the essence of this organic conquest/migration/inundation.

8. This is according to the Ethiopian manuscript "The Short Chronicles," cited in Muth 2007.

9. This is according to Leslau 1959 and my own observations.

10. These figures come from the Ethiopian National Census (2007), which was made using a sampling method. Hence it is likely that the numbers are either over- or underrepresentative, but there can be little doubt that a number of people in the Hararge region were still practicing the traditional religion. Moreover, numerous Oromo call themselves Muslim but still participate in traditional rituals.

11. This according to Waldron 1974.

12. See Caulk 1968 for an account of the events following the departure of the Egyptians.

13. According to Caulk 1987, Menelik offered Abdullahi terms under which he could retain his position of power in Harar and pay tribute to the Ethiopian state. Abdullahi refused.

14. A few Amhara and other Christians lived in Jugol. Some even converted to Islam and married Hararis.

15. The federal constitution also made provision for Hararis living anywhere in Ethiopia to vote for the government in their homeland. This was no shallow political gesture, as the population of Hararis in Addis and Dire Dawa is greater than that in Harar. This state of affairs had even led to a political party's emergence from the Harari population in Addis.

## CHAPTER 2

1. Phillips and Carillet 2006, 227. The Ras Hotels at the time were government-owned. I disagree with the authors' assessment that the Ras in Harar was overpriced and the staff dour. To me, the Ras was the best-value hotel in town, and the staff were some of the happiest and friendliest workers. The Ras Hotels have since been privatized and are a lot more expensive.

2. This story was later reiterated by several locals and the police commissioner. Given that they all said it happened in July 2009, I have no reason to suspect that it's an urban myth.

3. A local suggested to me that Sisay was trying to establish his own hyena-feeding operation for tourists. If this was the case, then it was certainly taking him a long time to do so. What's more, the government was actively preventing anyone other than the two official feeding places from hosting tourists. They didn't want a hyena-feeding operation on every street corner.

4. I offered to pay Sisay after I photographed him feeding the hyenas, but he refused the money.

5. This is from Shepard 1978, in which Shepard argues that the mind is a product of ecology. As such, animals hold the keys to the healthy maturation of humans and our ability to think and feel.

6. According to Shepard 1998, we need to recognize that we are products of the Pleistocene raised in the concrete and plastic environment of modernity. I'd argue that we have to some degree evolved beyond the Pleistocene, as many humans have the capacity to metabolize lactose, a tolerance for carbohydrate-rich diets, and a resistance to diseases such as smallpox and bubonic plague.

7. Shepard 1996, 282. This book has had a great influence on my thinking, even if I disagree on some points.

8. The playing field in Harar is reminiscent of Quammen's 2003 description of the town of Racadau in Romania, where people living in communist-era high-rise apartments went out at night to watch brown bears rummaging through garbage containers. The local

government in Racadau has made efforts to prevent bears from gaining access to the containers in the hope that they will stay "safely" in the forest.

9. Not a lot of houses in Harar had a water pipe connected, and even those that did had running water only on Saturdays, when people filled up the tanks that would supply them with water for the rest of the week.

10. This is an interesting etymology. Harari is a Semitic language, but *waraba* is a word of Cushitic origin. The Somali word for hyena is *warabo*, which I understand can also be taken to mean "newsman," but the Oromo *warabessa* has no such meaning.

11. One Harari interpreted this saying to mean that a hyena is always so hungry that he will eat anything, until he is left only with ash. After that, he's satisfied only because there's no other food. I found that when I asked the question of Hararis from Harar, they readily gave me the answer, but when I asked it of Hararis from Addis or the diaspora, it just puzzled them.

12. In Ethiopia it is customary to keep knocking even after the person inside has called out that they're coming to open the door.

CHAPTER 3

1. The honorific Aw translates as "father," indicating that Dudzo must have become one of the awliach, the saints, in Harar.

2. The French poet Arthur Rimbaud spent considerable time in Harar working for a French trading organization. The museum in Jugol has formed links with the French consulate in Ethiopia and secured funding from the French government for refurbishing the museum. Homosexuality is a crime in Ethiopia, but Rimbaud's homosexuality is overlooked in Harar and he is touted as an honorary Ethiopian.

3. From Pankhurst's description, it sounds like the hyena-feeding place was just past the current location of the Sofi dens, east of Jugol.

4. Kruuk 1968, 51.

5. Among Ethiopians it is customary to adopt a serious manner and count money very slowly in front of the payer.

6. Mulugayta also did this trick with men and older women, but mostly with pretty young girls.

7. Shoulder blades were the most prized of the bones and were often the subject of a lot of squabbling.

8. It was also not uncommon for tour guides to be boy toys for aging female tourists, who paid for their services over several nights.

9. Some guides specialized in French or Italian, but very few were fluent in whatever language they professed. Normally, they ignored or failed to understand tourists' questions. One tour guide in Harar was a mute who simply pointed out tourist attractions.

10. Compounding this unreliability was an extortionist who operated outside the Sofi feeding place and demanded a commission from the guides. Yusuf's wife asked me to do something about it, so I went to see the deputy police commissioner, Abdurrahman. He was extremely helpful and put an end to the extortionist's practices.

11. Ka Ea 2010b, 1. On arriving at the feeding place, the journalist was told by her guide that the tradition had begun two hundred years earlier, during a time of famine, and that the contemporary feeding was an extension of that practice (Ka Ea 2010a).

12. See Knight 2009. It's interesting that hyenas have such a strong aversion to humans, while animals like the albatross, who have had little experience of humans in evolutionary history, allow us to walk right up to them. This suggests that hyenas have evolved perceptual templates that allow humans to easily elicit fear reactions; humans may have something similar with regard to hyenas. This resonates with psychologist Martin Seligman's (1971) preparedness theory, which postulates that evolutionarily relevant, fear-inducing stimuli—such as spiders and snakes—are more likely to elicit fear responses in humans than stimuli such as guns and overloaded power outlets. See also Öhman and Mineka 2001.

CHAPTER 4

1. I say "relatively" because there are accounts of hyenas usurping alpha females and assuming the highest rank in the clan. There are also accounts of allies teaming up against higher-ranking hyenas.

2. For a summary of spotted hyena social organization and a comparison with other species of hyena, see Watts and Holekamp 2007.

3. The population estimate for spotted hyenas in Mills and Hofer's 1998 status survey and

conservation action plan is 27,800 to 48,200. This estimate was arrived at after a questionnaire survey was sent to various experts in the countries where hyenas were believed to be extant. Given that the maximum number of hyenas in Ethiopia is given as two thousand, while I estimate that there are two hundred around Harar alone, I think that this population estimate could be significantly low, at least in Ethiopia.

4. Tukwondilli was in fact outranked by Baby, so not only did he have to get the meat and escape before Dibbey arrived, he had to do so before Baby displaced him.

5. When hyenas are feeding, a loud alarm rattle from one causes the group to scatter and check for danger from a safe distance. Holekamp, Sakai, and Lundrigan (2007) cite anecdotal evidence to suggest that low-ranking hyenas use alarm calls deceptively to gain access to carcasses.

6. Buber actually speaks of the I-Thou relation; I use You here in place of Thou.

7. See Smuts 2001 on intersubjectivity between humans and animals, which describes the author's experiences with chimps, baboons, and her dog, Safi.

CHAPTER 5

1. The shrines that are said to provide porridge for hyenas are Aw Nugus, Amir Talhah, Aw Hakim, Aw Sofi, Aw Aboker, Aw Berkhadlee, Abdul Qadir Jailan, Aw Abdul, Aw Sayeed Ali, and Aw Qurabe Limay. The last three are right on the outskirts of Harar, Aw Hakim is a couple of hours' walk, and the rest are a considerable distance. According to Abdulmuheimen, there was once a family in Harar called "hyena porridge family" who used to feed porridge to hyenas within Jugol. The lineage ended after the battle of Chelenko.

2. The story of Emir Nur gives the hyenas a different motive for attacking the townspeople. In that version, the hyenas are concerned about the wall that he is building, which will shut them out of the town at night.

3. See Gibb 1996 and Zekaria 1991.

4. The museum collection included old Harari clothing, footwear, weapons, and manuscripts. Bookbinding was apparently once an art in Harar.

5. I was bemused to see the official typist for this government department typing with two fingers, like two finicky birds pecking away at the

keys. I guessed that when it came to landing a job as a typist in Harar, typing ability was not as important as family connections.

6. Even though the men and women celebrated in separate rooms, there were no restrictions on either sex's entering either of the rooms to greet people or pass on messages.

7. I used an infrared searchlight that I modified to bolt onto the tripod attachment of my camera. The setup was heavy and cumbersome, but I could film in complete darkness for up to fifty meters and pick up the reflections of hyenas' eyes for fifty meters beyond that. All the hyenas saw of it was a dim red glow coming from the searchlight.

8. Nureddin's opposition party lost the election.

9. To get a public service position in Harar, it certainly helps to be a member of the ruling party.

10. The video footage was pretty unimpressive. It was just a skittish hyena slurping butter from three mounds of porridge before trotting back into the forest. When I showed it to Nureddin and Hasanat, their disappointment was palpable. I'm sure they were expecting to see a handsome white hyena king testing porridge and signaling to his followers to join him and eat.

11. The festivities in Harar include a lot of porridge making and the Wirshato celebration (see Abdulnassir 2010 on the latter). Over the week leading up to Ashura and the Wirshato celebration, boys in Jugol prepare sticks and gourds. On the eve of Ashura they gather en masse and encourage people to throw old gourds for them to set upon and smash while singing the Wirshato song. There was recently a debate among the Hararis over whether the gourd smashing represented banishing alcohol from the town or whether it symbolized "smashing" corrupting influences from outside, although these are not necessarily mutually exclusive. Nowadays, Oromo and Amhara boys living in Jugol participate in the gourd smashing alongside Harari boys, which suggests that the purported symbolism is lost on those who just want to have a good time smashing gourds to bits in Harar.

12. The national television network apportions time slots to TV stations for numerous ethnic groups in Ethiopia. Harari TV airs nationally on Wednesday afternoons.

13. After I left Harar, the Tourism, Sport, and Culture Bureau contacted me and asked if I would contribute to its application to

UNESCO to have the Ashura event safeguarded as part of Harar's "intangible cultural heritage." I saw warning signs in the budget, which included promotional T-shirts, caps, and key chains. On the eve of Ashura in 2012, they installed floodlights at Aw Nugus to make it easier to film, and they put out five bowls of porridge to attract more hyenas. The result was a foregone conclusion: scores of curious people, no hyenas.

14. At the shrine of Aw Hakim, the murid solved the problem by holding the celebrations with the people on the eve of Ashura and putting out porridge for hyenas on the following night.

CHAPTER 6

1. I detailed these attitudes and beliefs in Baynes-Rock 2013b.
2. Waldron 1974 describes considerable gene flow between the Hararis and the Oromo; Leslau 1959 describes the shared origins of many Harari words—so why not ideas? From discussions with a few Somali people, I had the impression that they held a similar set of beliefs about hyenas. It could well be a regional thing.
3. Abbas called the same hyena Jalla, and this is the name I used in my blog. It wasn't unusual for father and son to give a hyena different names, as Abbas was normally in the house while Yusuf was feeding, and vice versa.
4. Abbas also referred to Kamareya by the Harari euphemism for a thief, *igi harir*, or "silken hands." This was because of his predilection for stealing food.
5. Of all the hyenas whose breathing I heard up close, only Kamareya wheezed like that.
6. The hyena hole opposite the Sofi feeding place was a main thoroughfare for people who wanted to pass through the wall between Argobberi and Suqutatberi. It was about 1.6 meters high.
7. The locals call these lanes *tcher uga* and use them only for drainage, waste dumping, and defecation. The others are known as *kachin uga* and it is these latter along which people travel.
8. Little was known about this aspect of hyena ecology before Kruuk's fieldwork in the 1960s.
9. See Tilson and Henschel 1986, which found clans of three, four, and five adults occupying territories of up to 816 square kilometers.
10. See Kruuk 1972 and 1975 and van Lawick-Goodall and van Lawick-Goodall 1970. I was greatly influenced by Jane Goodall's approach to the study of hyenas.
11. Wendy Binder and Blaire Van Valkenburgh (2000) conducted a study on the captive Berkeley clan. They found that while morphological growth measurements in hyenas reached a plateau at twenty months, bite force continued to increase until five years of age. They suggested that cubs' inefficiency at processing bones would favor later weaning times. Hyenas suckle for upward of eighteen months, which is a long time compared to other carnivores. Incidentally, one hyena in this study exerted a bite force of 4,500 newtons on the measuring instrument.
12. Smuts 2001, 295.
13. Compare the account in Despret 2004 of a horse named Clever Hans. While his handler maintained that Hans could give answers to math problems by tapping his hoof on the ground, the investigations of psychologist Oskar Pfungst found that Hans was in fact taking cues from the involuntary body language of the person expecting the answer. The phenomenon of responding to observer expectations is now known as the "Clever Hans effect."

CHAPTER 7

1. Sambusas are the Ethiopian equivalent of Indian samosas. Normally they're made with brown lentils, but some women make them with beef. At the time of my fieldwork, street vendors used frying pans on open fires or portable gas stoves, but these days they use electrical deep fryers with long extension cords, plugged into power outlets of nearby shops.
2. Marbles and knuckle jacks were popular games in the lanes of Harar. Toys were a rarity; a metal hoop with a wire to guide it as it spun along the road was about the best a kid could hope for.
3. This is from Sunan Abudawud (1:6; 41:5236). A similar conception is held in Malik's Muwatta (54:12:33).
4. It seems that jinn translate easily into traditional beliefs across Africa. See, for example, Aguilar 1995, and compare Bartels 1983. See also Thayer 1983; Boddy 1988; Bravmann 1977; Fadlalla 2005; and Ryan 2007.
5. Traditional Ethiopian stools have three legs and are carved from a single section of a tree trunk.

6. East and Hofer 1991, 640.

7. Theis et al. 2007.

8. See van Lawick-Goodall and van Lawick-Goodall 1970, 187-88, for an account of cubs seizing upon the gag-balls of a retching hyena.

9. Sausages known as *wakalim* are a Harari specialty, usually consumed only at weddings.

10. The Harar example of exorcism by hyena is quite mild in comparison to the Somali method, which involves locking a patient in a cage with a hyena. This must be traumatic for both parties. See http://www.bbc.co.uk /news/magazine-24539989.

11. According to local accounts, jinn are unable to cross bodies of water.

12. I did delete the recording. The text I've used here is not a verbatim transcription of the interview but an account taken from my written notes.

CHAPTER 8

1. This is, of course, a generalization. Hyenas also surround prey, trap prey in water, and catch prey in brief high-speed chases.

2. The two males were Belitti and his companion. It appeared that the Sofi hyenas knew these two and accepted their intermittent stays within the clan's territory.

3. I think Hadha Kamar was Willi's mother, as I often found these two together.

4. This was the night of March 29, 2010. I don't know if it is significant in terms of hyena playfulness, but on this and the following night, the moon was full.

5. It was impossible to see clearly these creatures darting about in the near darkness under the vast canopy of the tree. Consequently, I don't know what species they were, or whether these little flying things were birds or bats. There were, however, several weaver nests in the tree.

6. I think that Shepard would not draw the line at these two explanations of the function of play. He'd probably agree that predator/prey performances are also practice for small children, who are adapted to a world replete with child-snatching predators. In a Pleistocene environment, to which Shepard argues we are best adapted, chase-play and hide-and-seek games are essential practice. See Shepard 1996.

7. You could draw parallels here with sports, Carnivale, and all manner of activities where the status quo is suspended or even inverted.

8. The heat from Willi's leg was almost uncomfortable; such is the high body temperature of hyenas.

CHAPTER 9

1. Institutions in Ethiopia rarely have reception desks or signage indicating who is in what office and on which floor. Standard practice is to walk into the first office you find and get directions. Because of this, I imagine that office workers on the ground floor would be less productive and more irate than those on the floors above.

2. Foreigners and locals with enough money preferred Yemage Hospital. This place was built by a returned Harari expat from America. It was more clinic than hospital, but reassuringly clean.

3. A friend of mine working for a German NGO discovered this to his chagrin when someone hit his stationary vehicle in Addis. He was made to pay for the damage to both vehicles.

4. The Degue regime instituted a system of five levels of government: federal, state, zone, wereda, and kebele. Within East Hararge Zone in Oromia State, there are eighteen weredas, including Kombolcha Wereda. Within Kombolcha Wereda, there are twenty kebeles (one urban and nineteen rural), each with an administrative office. The population of a kebele in Kombolcha Wereda ranges from one thousand to twelve thousand people. It would not be unfair to suggest that Ethiopia is slightly overgoverned.

5. *Lantana camara* is widespread in the region. While it is an invasive species, farmers readily utilize the plants for hedgerows and boundaries. The name derives from *kolo*, which is a staple roasted barley, and *af*, meaning bird. In other words, the name suggests birds' fondness for the seeds.

6. It's extremely common for hyenas to bite the head when they attack humans. Given that they normally bite the undersides and hindquarters of ungulates, this suggests that the way they attack humans is adaptive.

7. Provided, that is, that governments can keep a lid on massive deforestation and habitat destruction.

8. This is the theme of Quammen's 2003 book, *Monster of God*. There is no answer to the conundrum, just accounts of displaced, powerless, fearful people and threatened predators.

9. The "EWCA Wildlife Policy and Strategies 2005" document uses the term "wildlife resources" thirteen times in its eleven pages.

CHAPTER 10

1. The local hotels mostly serve Harar Beer, brewed and bottled in a modern factory in Harar's New Town. No Hararis ever complained to me about the presence of a brewery in Harar.

2. The book *Notes from the Hyena's Belly* (Mezlekia 2000) is an autobiographical account of the author's time in Ethiopia during the Ethiopian / Somali war and the Dergue regime. The way he describes hyenas' agency and intentionality comes across as tongue in cheek but is in fact the way most people in the region conceive of hyenas.

3. Hyenas can cover a lot of ground quickly. In the course of his research in the Kalahari, Gus Mills (1990) recorded distances traveled by hyenas. One group of five traveled sixty-nine kilometers in under eleven hours. One male traveled 29.7 kilometers in just three hours.

4. The day after Election Day was more dangerous. The prime minister roused the public, accusing Human Rights Watch of interfering with Ethiopia's election process. Large crowds turned out in the streets of Harar to protest foreign interference, so I stayed indoors.

5. Waldron 1974, 270. It's very interesting to compare the old social order with the present, especially with regard to the Oromo.

6. Locals in Harar assert that some people are particularly prone to paranoia after chewing khat. Supporting this claim, the behavioral sensitization exhibited in laboratory animals to whom khat has been administered is strikingly similar to the development of psychosis and paranoia in humans administered psycho-stimulants. See Al-Habori 2005. The Oromo scholar Ezekiel Gebissa (2008) argues that the evidence for khat's negative effects is limited and that khat can provide an "economic life-line" in the region that outweighs the harms. See also Gebissa 2004.

CHAPTER 11

1. Minibuses in East Hararge (and all of Ethiopia, for that matter) are not only a danger to hyenas. In 2011, a minibus on its way to the Erer Valley plunged off the road about three kilometers outside Harar, killing thirty people.

2. Apparently, young boys in the suburb of Gunda Fera made biscuits laced with poison and left them out for hyenas, though it's not certain that that's what this juvenile ate.

3. From the accounts by Hans Kruuk and Jane Goodall, hyenas engage in cannibalism, though without the enthusiasm with which they devour other animal species. Hyenas have never been observed hunting other hyenas for the purpose of eating them.

4. Researchers Nicole Burgener and associates (2009) analyzed the chemical composition of eighty-four different hyenas' scented paste. The results are interesting and indicate that the scents also carry a badge of status. Thus lower-ranking hyenas like to anoint themselves with the paste of higher-ranking ones.

5. See Kruuk 1972 for details of hyena territorial behavior.

6. In Harar it is customary to discard any kind of packaging by throwing it on the ground. This even applies inside homes, where such things as candy wrappers are thrown directly on the floor, to be swept up later, usually by a young girl or servant.

7. I used this term in a special issue of the French journal *Social Science Information* that explored the topic "life in common" (Baynes-Rock 2013a).

CHAPTER 12

1. Like Hararis, Oromo people have neighborhood groups in which members contribute toward weddings and funerals. Naime's *afocha* contributed a canvas gazebo for the wedding, but it was used by the men for khat chewing, while the celebrations took place in the house.

2. Were-hyena is a neologism, analogous to werewolf, used to describe humans transformed into hyenas. In Africanist literature there are also accounts of were-jackals and were-leopards.

3. Mulugayta later found out that Sisay was feeding the Aboker hyenas. He threatened the butcher and made him stop. Sisay moved his butcher shop a few hundred meters away, and the hyenas no longer congregate at the playing field.

4. I am indebted to Dominique Lestel and colleagues for this: "When a singular animal adopts a singular trajectory within a group, its status becomes problematic, and its characterization as a member of a given species is undetermining" (2006, 171).

5. Known by the Amharic term *doro wot*, chicken curry in Ethiopia traditionally contains twelve pieces of chicken and three or four boiled eggs. The ability to ensure the correct number of chicken pieces in doro wot is one of many measures of a woman.

6. The hill behind the feeding place was an archaeologist's dream. It was in fact a refuse mound. At its eastern edge it was fifteen to twenty meters high, made up of year upon year of Harar's garbage. It also provided a nice view of the surrounding farmland.

7. Owing to Dire Dawa's proximity to Somaliland, security is extremely tight at the airport and in the town in general. The immigration office is also very wary of Somali people who attempt to get Ethiopian passports.

## CHAPTER 13

1. Arianna and I referred to the stone-throwing kids as the "Harar welcome."

2. See Holekamp et al. 1993. The authors suggest that a combination of low prey availability, increased competition for food, and the opening up of an adjacent territory were determining factors in the split. Still, not all low-ranking hyenas chose to emigrate, so the decision was not entirely determined by these environmental factors.

3. The authors suggest that the aggression toward the returnees could represent a general pattern of resistance to female immigrants, but never discount a hyena's memory. I doubt that a hyena would forget the scent of a clan mate after only an eight-month absence.

4. Ukhat is the Harari name for a staple more commonly known as *injera*. It is normally made from a local grain called *tef*, though it's often adulterated with maize or wheat flour. The flour is made into a batter, fermented, and then cooked on a ceramic hotplate to the consistency of a pancake with lots of bubbles. While injera in restaurants is quite thin, that made by rural people is a lot thicker, supposedly reflecting the greater appetite of people who work the land.

5. As a general rule, male hyenas emigrate, but there are always exceptions. Sons of high-ranking mothers especially have little motivation to join another clan, where their rank will plummet to the bottom. In the case of Tukwondilli, Yusuf maintained that he was born into the Sofi clan. He apparently went away for up to three months at a time but always returned. I suspect that his absences were visits to other clans to find mates, and that his returns were for feeding opportunities and possibly infatuation with Dibbey.

6. Lawrence Frank and associates (1990) at UC Berkeley developed a noninvasive means of determining the sex of hyenas (older than three months) based on phallic morphology. Whereas the phallic glans in female hyenas is symmetrical in profile and club-shaped, that of the males is asymmetrical, spade-shaped, and pointed at the end. Sex can often also be determined on the basis of the pronounced testes in the scrotum and upward-curving ventrum of males.

7. Hans Kruuk (1972) used the term "greeting ceremony" for this behavior and wondered whether the enlarged genitalia of the females were adapted for this purpose.

8. Standard practice in the Maasai Mara is to follow hyenas from 5:30 to 9:00 A.M. and from 5:00 to 8:00 P.M. In study periods between 2002 and 2004, Joe Kolowski used infrared spotlights and night-vision goggles to follow hyenas day and night. Using composites of hyena-following periods, he and his associates found that the vast majority of hyena activity occurred during the night.

9. The Michigan State University project has since recorded a case of adoption among the Talek West hyenas. After a poisoning incident in April 2014, an orphaned cub, Cyberman, was observed following and consistently feeding from an unrelated female.

10. The poisoning incident of April 2014 was estimated to have killed at least nineteen hyenas, plus some dependent cubs who subsequently starved to death.

11. According to Kolowski and Holekamp 2007, lower-ranked hyenas favored the refuse pit, but this wasn't a general rule. One low-ranking hyena was rarely seen at the place, while one high-ranking female frequented it regularly.

## CHAPTER 14

1. Apparently, Abdulmuheimen documented more than two hundred named elements of a Harari house.

2. In Harari, "Sari be zal waraba, afet qum zo be, at zo, rizq zo, yafach'ma gub zo be, ya gibba."

3. Whereas Kevin Theis and colleagues (2007) claimed that these "type T" terminal whoops occurred exclusively at the end of whooping bouts by Mara hyenas, Marion East and

Heribert Hofer (1991) found that this was not always the case with adults in Ngorongoro. In my experience in Harar, terminal whoops always came at the end of a bout.

CHAPTER 15

1. A habuka is a fruit that looks like a globular chayote and tastes like a very tart apple, yet very bland in terms of flavor.

2. The results were published by Christine Drea and Alissa Carter (2009). Anyone familiar with hyenas and the determination with which they can find food would be unsurprised that they solved the problem so easily.

3. While the young hyenas used in the study were all raised in captivity, they were descendants of cubs originally taken from the Maasai Mara. From time to time, cubs of Berkeley hyenas are given to zoos and new cubs are brought from the wild. Over the course of the Berkeley project, male and female cubs have been gonadectomized and females administered with antiandrogens to see how their genitals might develop differently from the hyena "norm." Wilson 2003 argues that the methods and theories underpinning the Berkeley studies perpetuate hegemonies of normalcy. According to Wilson, the study was in fact set up to warn of the "consequences of deviation from sex/gender norms" (Wilson 2003, 757). The project coordinator, Stephen Glickman, published a paper about the history of maligned hyenas, describing how the producers of *The Lion King* requested that Disney Studios' artists be allowed to study the Berkeley hyenas so as to get the artwork right. Glickman laid down a condition, that the hyenas in the film be portrayed positively, to which the producers agreed. The rest is movie history. Glickman also admitted to having developed a bond with the hyenas; in the published paper there is a photograph of Krista Glickman hugging an adult female in one of the holding pens. Project funding has always been an issue, and there has long been talk of the facility being shut down and the hyenas euthanized or given to zoos. This has prompted Kansas State University's Tona Melgarejo to launch a fundraising drive seeking US$150,000 to maintain the colony for one more year, until a permanent endowment can be established. But even if the dollars roll in, animal rights protesters threaten the colony. After five voles in a UC research facility died

from neglect and dehydration, protesters demanded that all animal research at Berkeley be shut down.

4. Testing for humanlike reasoning in animals is what Gary Francione (2005) calls the "similar minds approach." He argues that this approach reinforces distinctions between humans and animals by prioritizing sentience over other qualities.

5. The term "Disneyish" comes from Paul Shepard (1996), who decries the idea of fraternization with animals, suggesting that it is symptomatic of the neurosis of humans in industrialized societies who are disconnected from both animals and their own place among Others. For Shepard, re-creating a peaceable kingdom on earth was but a step toward projecting the domestic world onto wild nature. The outcome he foresaw was the imposition of human "moral eminence" and a greater disconnection of humans from animals.

6. Geographer Daniel Gade (2006) argues that the persistence of hyenas in the Horn of Africa at once complements human practices and is a vestige of ancient ecological relations.

7. I was in fact directed to it by an anonymous reviewer of my original manuscript.

8. Donna Haraway is influential in my thinking here. My discussion of human and hyena becoming like each other is a juxtaposition of her ideas about humans and dogs. Where there is an evolutionary history of dogs and humans going toward each other, evolving a shared understanding, a co-dependence, and even, as Haraway says, a sharing of DNA, a human becoming doglike is simultaneously becoming humanlike. It's the realization of a potentiality. The converse applies with humans and hyenas.

9. Kohn 2007 noted this concern among the Amazonian Runa; Willerslev 2007 also noted it among Yukaghir hunters in Siberia. Willerslev's thinking is influential with regard to mimesis as a way of apprehending the subjectivities of nonhuman Others. Where I diverge is on the point of personhood. Willerslev suggests that in mimicking nonhumans one risks losing one's personhood, but this position is based on the view that nonhuman Others are not persons. I argue that one's personhood is only at stake in that it is transformed by going toward the subjectivity of the Other.

# BIBLIOGRAPHY

Abdulnassir, A. 2010. "Wirshato: The Gourd-Smashing Ceremony." *African Study Monographs* 41:149–51.

Aguilar, M. I. 1995. "African Conversion from a World Religion: Religious Diversification by the Waso Boorana in Kenya." *Africa: Journal of the International African Institute* 65, no. 4: 525–44.

Al-Habori, M. 2005. "The Potential Adverse Effects of Habitual Use of *Catha edulis* (Khat)." *Expert Opinion on Drug Safety* 4, no. 6: 1145–54.

Arribas, A., and P. Palmqvist. 1999. "On the Ecological Connection Between Sabre-Tooths and Hominids: Faunal Dispersal Events in the Lower Pleistocene and a Review of the Evidence for the First Human Arrival in Europe." *Journal of Archaeological Science* 26:571–85.

Austen, R. A. 1993. "The Moral Economy of Witchcraft: An Essay in Comparative History." In *Modernity and Its Malcontents: Ritual and Power in Postcolonial Africa*, ed. J. Comaroff and J. Comaroff, 89–110. Chicago: University of Chicago Press.

Balestra, F. A. 1962. "The Man-Eating Hyenas of Mlanje." *African Wildlife* 16:25–27.

Barry, J. C. 1987. "Large Carnivores (Canidae, Hyaenidae, Felidae) from Laetoli." In *Laetoli: A Pliocene Site in Northern Tanzania*, ed. M. D. Leakey and J. M. Harris, 235–58. Oxford: Clarendon Press.

Bartels, L. 1983. *Oromo Religion: Myths and Rites of the Western Oromo of Ethiopia—An Attempt to Understand*. Berlin: Dietrich Reimer Verlag.

Baynes-Rock, M. 2012. "Hyenas Like Us: Social Relations with an Urban Carnivore in Harar, Ethiopia." PhD diss., Macquarie University, Sydney.

———. 2013a. "Life and Death in the Multispecies Commons." *Social Science Information* 52, no. 2: 210–27.

———. 2013b. "Local Tolerance of Hyena Attacks in East Hararge Region, Ethiopia." *Anthrozoös* 26, no. 3 (2013): 421–33.

Beidelman, T. O. 1963. "Witchcraft in Ukaguru." In *Witchcraft and Sorcery in East Africa*, ed. J. Middleton and E. H. Winter, 57–98. London: Routledge and Kegan Paul.

——. 1980. "The Moral Imagination of the Kaguru: Some Thoughts on Tricksters, Translation, and Comparative Analysis." *American Ethnologist* 7, no. 1: 27–42.

Binder, W. J., and B. Van Valkenburgh. 2000. "Development of Bite Strength and Feeding Behaviour in Juvenile Spotted Hyenas (*Crocuta crocuta*)." *Journal of Zoology* 252:273–83.

Binford, L. R. 1985. "Human Ancestors: Changing Views of Their Behaviour." *Journal of Anthropological Archaeology* 4:292–327.

Blumenschine, R. J. 1987. "Characteristics of an Early Hominid Scavenging Niche." *Current Anthropology* 28, no. 4: 383–94.

Boaz, N. T., R. L. Ciochon, Q. Xu, and J. Liu. 2000. "Large Mammalian Carnivores as a Taphonomic Factor in the Bone Accumulation at Zhoukoudian." *Acta Anthropologica Sinica* 19:224–34.

Boddy, J. 1988. "Spirits and Selves in Northern Sudan: The Cultural Therapeutics of Possession and Trance." *American Ethnologist* 15, no. 1: 4–27.

Bravmann, R. A. G.-G. 1977. "Making the Jinn Manifest." *African Arts* 10, no. 3: 46–52.

Brottman, M. 2012. *Hyena*. London: Reaktion Books.

Bunn, H. T. 1981. "Archaeological Evidence for Meat-Eating by Plio-Pleistocene Hominids from Koobi Fora and Olduvai Gorge." *Nature* 291:574–77.

Burgener, N., M. Dehnhard, H. Hofer, and M. L. East. 2009. "Does Anal Gland Scent Signal Identity in the Spotted Hyaena?" *Behaviour* 77:707–15.

Burton, R. F. 1856. *First Footsteps in East Africa*. London: Longman, Brown, Green and Longmans.

Caulk, R. A. 1968. "Harar in the Nineteenth Century and the Loss of Its Independence." In Caulk, "A Paper Prepared for the Interdisciplinary Seminar of the Faculties of Arts and Education," 1–15. Addis Ababa University.

——. 1987. "The Occupation of Harar: January 1887." *Journal of Ethiopian Studies* 9, no. 2: 1–20.

Central Statistical Agency. 2007. "Statistical Tables for the 2007 Population and Housing Census of Ethiopia (Harari Region)." Addis Ababa: Population Census Commission.

Chauvet, J. M., E. B. Deschamps, and C. Hillaire. 1996. *Dawn of Art: The Chauvet Cave; The Oldest Known Paintings in the World*. New York: Harry N. Abrams.

Ciekawy, D., and P. Geschiere. 1998. "Containing Witchcraft: Conflicting Scenarios in Postcolonial Africa." *African Studies Review* 41, no. 3: 1–14.

Despret, V. 2004. "The Body We Care For: Figures of Anthropo-Zoo-Genesis." *Body and Society* 10, nos. 2–3: 111–34.

Drea, C. M., and A. N. Carter. 2009. "Cooperative Problem Solving in a Social Carnivore." *Animal Behaviour* 78:967–77.

East, M. L., and H. Hofer. 1991. "Loud Calling in a Female-Dominated Mammalian Society: I. Structure and Composition of Whooping Bouts of Spotted Hyaenas, *Crocuta crocuta*." *Animal Behaviour* 42:637–49.

Fadlalla, A. H. 2005. "Modest Women, Deceptive *Jinn*: Identity, Alterity, and Disease in Eastern Sudan." *Identities: Global Studies in Culture and Power* 12:143–74.

Francione, G. 2005. "You Hypocrites!" *New Scientist*, June 4, 51–53.

Frank, L. G., S. E. Glickman, and I. Powch. 1990. "Sexual Dimorphism in the Spotted Hyaena (*Crocuta crocuta*)." *Journal of Zoology* 221:308–13.

Gabunia, L., A. Vekua, D. Lordkipanidze, R. Ferring, A. Justus, G. Maisuradze, A. Mouskhelishvili, et al. 2000. "Current Research on the Hominid Site of Dmanisi." *Eraul* 92:13–27.

Gade, D. W. 2006. "Hyenas and Humans in the Horn of Africa." *Geographical Review* 96, no. 4: 609–32.

Garrard, A. N. 1984. "Community Ecology and Pleistocene Extinctions in the Levant." In *Hominid Evolution and Community Ecology: Prehistoric Human Adaptation in Biological Perspective*, ed. R. Foley, 260–77. London: Academic Press.

Gebissa, E. 2004. *Leaf of Allah: Khat and Agricultural Transformation in Harerge, Ethiopia, 1875–1991*. Oxford: James Currey.

———. 2008. "Scourge of Life or an Economic Lifeline? Public Discourses on Khat (*Catha edulis*) in Ethiopia." *Substance Use and Misuse* 43:784–802.

Gibb, C. C. T. 1996. "In the City of Saints: Religion, Politics, and Gender in Harar, Ethiopia." PhD diss., University of Oxford, Oxford.

Gibbons, A. 2005. "Once-Balmy Climate Lured Humans to England Early." *Science* 308:490.

Gibert, J., D. Campillo, V. Eisenmann, E. García-Olivares, A. Malgosa, D. A. Roe, M. J. Walker, et al. 1999. "Spanish Late Pliocene and Early Pleistocene Hominid, Palaeolithic, and Faunal Finds from Orce (Granada) and Cueva Victoria (Murcia)." *Human Evolution* 14, nos. 1–2: 29–46.

Glickman, S. E. 1995. "The Spotted Hyena from Aristotle to the Lion King: Reputation Is Everything." *Social Research* 62, no. 3: 501–37.

Goldman, M. J., J. Roque de Pinho, and J. Perry. 2010. "Maintaining Complex Relations with Large Cats: Maasai and Lions in Kenya and Tanzania." *Human Dimensions of Wildlife* 15:332–46.

Gottlieb, A. 1989. "Hyenas and Heteroglossia: Myth and Ritual Among the Beng of Côte d'Ivoire." *American Ethnologist* 16, no. 3: 487–501.

Gray, R. F. 1963. "Some Structural Aspects of Mbugwe Witchcraft." In *Witchcraft and Sorcery in East Africa*, ed. J. Middleton and E. H. Winter, 143–74. London: Routledge and Kegan Paul.

Haraway, D. J. 2003. *The Companion Species Manifesto: Dogs, People, and Significant Otherness*. Chicago: Prickly Paradigm Press.

Hemingway, E. 1935. *Green Hills of Africa*. New York: Scribner.

Holekamp, K. E., J. O. Ogutu, H. T. Dublin, L. G. Frank, and L. Smale. 1993. "Fission of a Spotted Hyena Clan: Consequences of Prolonged Female Absenteeism and Causes of Female Emigration." *Ethology* 93:285–99.

Holekamp, K. E., S. T. Sakai, and B. L. Lundrigan. 2007. "The Spotted Hyena (*Crocuta crocuta*) as a Model System for the Study of the Evolution of Intelligence." *Journal of Mammalogy* 88, no. 3: 545–54.

Johnstone, H. B. 1902. "Notes on the Customs of the Tribes Occupying Mombasa Sub-District, British East Africa." *Journal of the Anthropological Institute of Great Britain and Ireland* 32:263–72.

Ka Ea, L. 2010a. "The Beauty and the Beasts of Harar: Part I." *Malaysian Insider*, February 6.

———. 2010b. "The Beauty and the Beasts of Harar: Part II." *Malaysian Insider*, February 23.

Knight, J. 2009. "Making Wildlife Viewable: Habituation and Attraction." *Society and Animals* 17:167–84.

Kohn, E. 2007. "How Dogs Dream: Amazonian Natures and the Politics of Transpecies Engagement." *American Ethnologist* 34, no. 1: 3–24.

Kolowski, J. M., and K. E. Holekamp. 2007. "Effects of an Open Refuse Pit on Space Use Patterns of Spotted Hyenas." *African Journal of Ecology* 46:341–49.

Kolowski, J. M., D. Katan, K. R. Theis, and K. E. Holekamp. 2007. "Daily Patterns of Activity in the Spotted Hyena." *Journal of Mammalogy* 88, no. 4: 1017–28.

Kruuk, H. 1968. "Hyenas: The Hunters Nobody Knows." *National Geographic* 134:44–57.

————. 1972. *The Spotted Hyena: A Study of Predation and Social Behavior*. Chicago: University of Chicago Press.

————. 1975. *Hyaena*. London: Oxford University Press.

————. 2002. *Hunter and Hunted: Relationships Between Carnivores and People*. Cambridge: Cambridge University Press.

Kurtén, B. 1968. *Pleistocene Mammals of Europe*. London: Weidenfeld and Nicolson.

Leakey, L. S. B. 1977. *The Southern Kikuyu Before 1903*. Vol. 1. London: Academic Press.

Leakey, M. D., and R. L. Hay. 1979. "Pliocene Footprints in the Laetolil Beds at Laetoli, Northern Tanzania." *Nature* 278:317–23.

Leslau, W. 1959. "A Preliminary Description of Argobba." *Annales d'Éthiopie* 3:251–73.

————. 1965. *Harari*. Vol. 1 of *Ethiopians Speak: Studies in Cultural Background*. Berkeley: University of California Press.

Lestel, D., F. Brunois, and F. Gaunet. 2006. "Etho-Ethnology and Ethno-Ethology." *Social Science Information* 45:155–77.

Levine, D. N. 1974. *Greater Ethiopia: The Evolution of a Multiethnic Society*. Chicago: University of Chicago Press.

Littman, E. 1936. "Harar." In *Encyclopedia of Islam*, 3:263–64. London, 1960.

Long, R. C. 1973. "A List with Notes of the Mammals of the Nsanje (Port Herald) District, Malawi." *Society of Malawi Journal* 26, no. 1: 60–77.

Louchart, A., H. Wesselman, R. J. Blumenschine, L. J. Hlusko, J. K. Njau, M. T. Black, M. Asnake, and T. D. White. 2009. "Taphonomic, Avian, and Small-Vertebrate Indicators of *Ardipithecus ramidus* Habitat." *Science* 326:66.

Marshall, L. 1976. *The !Kung of the Nyae Nyae*. Cambridge: Harvard University Press.

Marshall Thomas, E. 1993. *The Hidden Life of Dogs*. London: Phoenix.

Mezlekia, N. 2000. *Notes from the Hyena's Belly: An Ethiopian Boyhood*. New York: Picador.

Mills, G., and H. Hofer. 1998. *Hyaenas: Status Survey and Conservation Action Plan*. Gland, Switzerland: IUCN.

Mills, M. G. L. 1990. *Kalahari Hyaenas: Comparative Behavioural Ecology of Two Species*. London: Unwin Hyman.

Morris, B. 2000a. *Animals and Ancestors: An Ethnography*. Oxford: Berg.

————. 2000b. "Wildlife Depredations in Malawi: The Historical Dimension." In *Natural Enemies: People-Wildlife Conflicts in Anthropological Perspective*, ed. J. Knight, 36–49. London: Routledge.

Morwood, M. J., and W. L. Jungers. 2009. "Conclusions: Implications of the Liang Bua Excavations for Hominin Evolution and Biogeography." *Journal of Human Evolution* 57, no. 5: 640–48.

Muth, F. C. 2007. "Nur b. Mugahid." In *Encyclopedia Aethiopica*, ed. S. Uhlig, 2:1209–10. Wiesbaden: Harrassowitz.

Nelson, R. 1993. "Searching for the Lost Arrow: Physical and Spiritual Ecology in the Hunter's World." In *The Biophilia Hypothesis*, ed. S. R. Kellert and E. O. Wilson, 202–28. Washington, D.C: Island Press.

Niehaus, I. 1995. "Witches of the Transvaal Lowveld and Their Familiars: Conceptions of Duality, Power, and Desire (Sorciers de la région du lowveld au Transvaal)." *Cahiers d'Études Africaines* 35, nos. 138–39: 513–40.

Noske, B. 1993. "The Animal Question in Anthropology: A Commentary." *Society and Animals* 1, no. 2: 185–90.

Öhman, A., and S. Mineka. 2001. "Fear, Phobias, and Preparedness: Toward an Evolved Module of Fear and Fear Learning." *Psychological Review* 108:483–522.

Pankhurst, S. 1958. "Dinner Party with the Hyenas." *Ethiopia Observer*, March, 74–75.

Paulitschke, P. 1888. *Harar: Forschungsreise nach den Somâl und Gallaländern, Ost-Afrikas*. Leipzig: F. A. Brockhaus.

Phillips, M., and J. B. Carillet. 2006. *Ethiopia and Eritrea*. English ed. Melbourne: Lonely Planet.

Potts, R., and P. Shipman. 1981. "Cutmarks Made by Stone Tools on Bones from Olduvai Gorge, Tanzania." *Nature* 291:577–80.

Quammen, D. 2003. *Monster of God: The Man-Eating Predator in the Jungles of History and the Mind*. New York: W. W. Norton.

Rabinovich, R. 2002. "Man Versus Carnivores in the Middle-Upper Paleolithic of the Southern Levant." In *Archaeozoology of the Near East V: Proceedings of the Fifth International Symposium on the Archaeozoology of Southwestern Asia and Adjacent Areas*, ed. H. Buitenhuis, A. M. Choyke, M. Mashkour, and A. H. Al-Shiyab, 28–45. Groningen: ARC.

Routledge, W. S., and K. Routledge. 1968. *With a Prehistoric People: The Akikuyu of British East Africa*. London: Frank Cass.

Rutherford, B. 1999. "To Find an African Witch: Anthropology, Modernity, and Witch-Finding in North-West Zimbabwe." *Critique of Anthropology* 19, no. 1: 89–109.

Ryan, J. 2007. "Going 'Walli' and Having 'Jinni': Exploring Somali Expressions of Psychological Distress and Approaches to Treatment." PhD diss., University of Waikato, Hamilton, New Zealand.

Sapir, J. D. 1981. "Leper, Hyena, and Blacksmith in Kujamaat Diola Thought." *American Ethnologist* 8, no. 3: 526–43.

Sayers, G. F. 1930. *The Handbook of Tanganyika*. London: Macmillan.

Schepartz, L. A., S. Miller-Antonio, and D. A. Bakken. 2000. "Upland Resources and the Early Palaeolithic Occupation of Southern China, Vietnam, Laos, Thailand, and Burma." *World Archaeology* 32, no. 1: 1–13.

Seligman, M. E. P. 1971. "Phobias and Preparedness." *Behavior Therapy* 2:307–20.

Shepard, P. 1978. *Thinking Animals: Animals and the Development of Human Intelligence*. Athens: University of Georgia Press.

———. 1996. *The Others: How Animals Made Us Human*. Washington, D.C.: Island Press.

———. 1998. *Coming Home to the Pleistocene*. Washington, D.C.: Island Press.

Shipman, P. 1983. "Early Hominid Lifestyle: Hunting and Gathering or Foraging and Scavenging?" In *Animals and Archaeology: Hunters and Their Prey*, ed. J. Clutton-Brock and C. Grigson, 31–49. BAR International Series 163. Oxford: British Archaeological Reports.

Smuts, B. 2001. "Encounters with Animal Minds." *Journal of Consciousness Studies* 8, nos. 5–7: 293–309.

Stiner, M. C. 2004. "Comparative Ecology and Taphonomy of Spotted Hyenas, Humans, and Wolves in Pleistocene Italy." *Revue de Paléobiologie, Genève*, 23, no. 2: 771–85.

Thayer, J. S. 1983. "Nature, Culture, and the Supernatural Among the Susu." *American Ethnologist* 10, no. 1: 116–32.

Theis, K. R., K. M. Greene, S. R. Benson-Amram, and K. E. Holekamp. 2007. "Sources of Variation in the Long-Distance Vocalizations of Spotted Hyenas." *Behaviour* 144:557–84.

Tilson, R. L., and J. R. Henschel. 1986. "Spatial Arrangement of Spotted Hyena Groups in a Desert Environment, Namibia." *African Journal of Ecology* 24:173–80.

Trimingham, J. S. 1965. *Islam in Ethiopia*. London: Frank Cass.

van Lawick-Goodall, H., and J. van Lawick-Goodall. 1970. *Innocent Killers*. London: Collins.

Waldron, S. R. 1974. "Social Organization and Social Control in the Walled City of Harar, Ethiopia." PhD diss., Columbia University.

Watts, H. E., and K. E. Holekamp. 2007. "Hyena Societies." *Current Biology* 17, no. 16: 657–60.

Willerslev, R. 2007. *Soul Hunters: Hunting, Animism, and Personhood Among the Siberian Yukaghirs*. Berkeley: University of California Press.

Wilson, A. 2003. "Sexing the Hyena: Intraspecies Readings of the Female Phallus." *Signs: Journal of Women and Culture in Society* 28, no. 3: 755–90.

Zekaria, A. 1991. "Hyena Porridge: Ethnographic Filming in the City of Harar." *Sociology Ethnology Bulletin* 1, no. 1: 86–88.

# INDEX